ATOMIC THEORY

ATOMIC THEORY

AN ELEMENTARY EXPOSITION

BY

ARTHUR HAAS, PH.D.

PROFESSOR OF PHYSICS IN THE UNIVERSITY OF VIENNA

TRANSLATED BY

T. VERSCHOYLE, M.C., PH.D., A.R.C.S.

LONDON

CONSTABLE & COMPANY LTD

10–12 ORANGE STREET W.C. 2

1927

Printed in Great Britain by The Whitefriars Press, Ltd.,
London and Tonbridge.

TO

FREDERICK GEORGE DONNAN

C.B.E., F.R.S.,

PROFESSOR OF INORGANIC AND

PHYSICAL CHEMISTRY

IN THE

UNIVERSITY OF LONDON

EXTRACT FROM PREFACE TO THE GERMAN EDITION

THE present book has been compiled from lectures which I delivered at the University of Vienna. The object of these lectures was to explain the fundamental ideas and the most important consequences of modern atomic theory in a clear and elementary fashion, avoiding the use of higher mathematics. Since only a " school " knowledge of physics is assumed on the part of the reader I hope that this book may be of use, not merely to physicists, but also to chemists and even, perhaps, to those who, without possessing any great knowledge of science, wish to learn something of atomic theory. . . .

ARTHUR HAAS.

TRANSLATOR'S NOTE

THE text of the " Atomic Theory " is practically identical with that of Part III (of Volume II) of the author's " Introduction to Theoretical Physics " ; only minor alterations have been made. The object in publishing this part separately is explained by Professor Haas's preface, for it is intended to appeal chiefly to those who do not desire to buy the two volumes of the larger work, as covering more ground than they wish to explore, but who want a clear and readable exposition of the rapid progress in modern atomic theory. To such readers this book may be confidently recommended, for Professor Haas has a peculiar and very happy gift of exposition ; any obscurity or inaccuracy of expression must be attributed to the translator.

The two years that have elapsed between the appearance of the German and the present English edition have seen the accomplishment of much new work in atomic theory. It is, indeed, almost impossible to keep pace with the development of the theory of spectra, but Professor Haas has very kindly corrected old and added new matter, so that the English edition should be as up to date as can reasonably be expected. For this, and for the continual help and valuable advice which Professor Haas has given, the translator is deeply indebted, as he is also to Dr. R. W. Lawson, of the University of Sheffield, for his careful revision of the manuscript, and numerous suggestions for improvement. He owes also many grateful acknowledgements to both the publishers and the printers, for whom nothing has been too much trouble.

T. VERSCHOYLE.

PREFATORY NOTE

The quantities encountered in atomic physics are given both in *absolute* and in *technical units*. The fundamental units of the absolute system of measurement are the gram, the centimetre, and the second. From them is derived as unit of force the *dyne*, *i.e.*, the force which imparts an increase in velocity of 1 cm./sec. to unit-mass (1 gm.) in unit-time (1 sec.); a dyne is equivalent to the weight of 1·0198 mgm. (at a latitude of 45°). The work performed by a force of 1 dyne along a path 1 cm. long is defined as an *erg*, and represents the unit of energy. $4{\cdot}186 \times 10^7$ ergs are equivalent to the unit of heat, the gram-calorie, which is the heat required to raise the temperature of 1 gm. of water from 15° to 16°.

The *electrostatic unit of electricity* is that quantity which, in accordance with Coulomb's Law, exerts a force of 1 dyne upon an equal quantity at a distance of 1 cm. If the force acting upon an arbitrary quantity of electricity at any point in an electric field be divided by that quantity, the quotient represents the *electric strength of field* at the point in question. The *unit of magnetism* and magnetic strength of field are defined analogously to the electrostatic unit and electric field-strength.

The absolute unit of *electric potential* (in so-called electrostatic measure) is given by a potential difference such that, on moving through it, 1 electrostatic unit performs work equal to 1 erg in amount. The three-hundredth part of this absolute unit of potential represents the technical unit of electric potential, or *volt*. The absolute unit of current (in electrostatic measure) is represented by a rate of flow of 1 electrostatic unit in a second. The technical unit, or *ampère*, is equal to 3×10^9 absolute units.

In spectroscopy, the micron and the Ångström unit are used as units of length: 1μ is equal to 10^{-4} cm., and 1 Å.U. to 10^{-8} cm.

CONTENTS

CHAPTER I

The Elementary Quanta

CHAPTER II

Theory of the Hydrogen Atom

CHAPTER III

Röntgen Rays

CHAPTER IV

Theory of the Elements

CHAPTER V

General Theory of Spectra and of Atomic Structure

APPENDIX

ATOMIC THEORY

CHAPTER I

THE ELEMENTARY QUANTA

§ 1. The Elementary Quantum of Electricity.

IN the whole of theoretical physics there is no older conception than that of the *individual structure of matter*, according to which matter consists of minute, invisible, *individual particles*. This idea originated long ago with the Greek philosopher *Democritus*, who gave the name *atoms* to the ultimate invisible particles into which matter could be resolved. This term has also been retained in modern science for the *smallest particles of the chemical elements*, and it is to the building up of atoms to form so-called molecules that the newer chemistry, founded by *Dalton*, ascribes the multiplicity of chemical compounds.

Numerous physical phenomena, however, have clearly shown that *atoms* must undoubtedly *contain electrical charges*. Above all, the fact of *electrolysis* demonstrates that atoms are influenced by electrical forces. Moreover, after the establishment of the electromagnetic theory of light, the long known phenomenon of the *dispersion of light* effected by matter could not be explained without the assumption of a mutual action between electromagnetic waves and electrical charges contained in the atoms of matter. The hypothesis of an atomistic structure of matter thus necessarily led to the hypothesis of an *atomistic structure of electricity*, and hence to the conception of an *elementary quantum of electrical charge* (or elementary charge). Not only theoretical considerations, however, have given rise to this fundamental assumption of modern atomic theory, for a *direct empirical confirmation* is to be found in some *experiments of Millikan*, which have, moreover, rendered possible a direct and extremely accurate determination of the elementary quantum of electricity.

Millikan's experiments consisted in the observation with a microscope of *minute, individual particles of matter* carrying *feeble electric charges*, and moving under the two-fold influence of gravity and of a field directed vertically upwards.

This method of so-called "*individual observation*"[1] was devised in 1909 by *Ehrenhaft* and a little later also by Millikan, and was further developed and perfected in the following years.[2]

Millikan worked with *oil droplets* with a diameter of about 10^{-4} cm. The advantage of oil droplets, as compared with particles of other substances, is that they do not change in size owing to evaporation. By means of a spray-diffuser a fine mist of oil is blown into a chamber, and through a small hole in the bottom of the latter a droplet is from time to time admitted into a condenser placed beneath the chamber, the condenser consisting of two parallel horizontal plates. In consequence of the friction due to the spraying, these droplets are almost always electrically charged. The plates of the condenser can be connected through a commutator to a battery of accumulators giving some 10,000 volts. The particle in the interior of the condenser is illuminated by a strong source of light placed at one side, and it is at the same time observed by a microscope through a window in the condenser; it then appears as a bright point on a black background, although its true shape cannot be recognized. When the electric field is cut out, a particle of this kind slowly falls with a constant velocity, in consequence of its own weight and the strong frictional resistance. When, however, the battery is switched on, the particle slowly rises, again with a constant velocity, provided that the electric force is not much greater than that of gravity. (With the given size of droplet and potential difference the conditions are just right.) By always switching on or off at the correct moment, it is possible to keep a droplet moving up and down between the two plates of the condenser as long as is required.

On account of the large friction, the *velocity* of the particle in both the upward and the downward movement is *proportional to the effective force*, as follows both from mechanical considerations and by direct experimental confirmation. Denoting, therefore, the velocity of the downward motion by v_1 and that of the upward motion by v_2, the mass of the particle by m and its charge by Q, the acceleration due to

[1] The essential novelty of this method lay in the fact that it dealt with *individual* particles, whereas earlier methods had always dealt with the *mean value* of the charges on a *large number* of particles.

[2] A collected report of Millikan's researches is to be found in his book, "*The Electron, its Isolation and Measurement*" (Chicago, 1917). His first paper was published in August, 1909 (in the reports of the British Association Meeting at Winnipeg); Ehrenhaft's first communication appeared in March, 1909 (in the *Anzeiger der Wiener Akademie*).

gravity by g, and the electric strength of field (*i.e.*, the quotient of the potential difference and the distance between the plates) by E, we must have the following relation :—

$$(1) \qquad \frac{v_2}{v_1} = \frac{QE - mg}{mg}.$$

The velocity is easily determined by measuring the time taken by the particle in traversing the interval between two cross-threads in the observing microscope. Table I gives the values in seconds found in one experiment by Millikan for the times (t_1) required for the downward motion, and for the times (t_2) required for the upward motion between the cross-threads ; in this case the distance between the threads corresponded to an actual interval of rise or fall of 0·5222 cm.

TABLE I

t_1	t_2	t_1	t_2
13·6	12·5	13·8	34·6
13·8	12·4	13·7	34·8
13·4	21·8	13·8	16·0
13·4	34·8	13·6	34·8
13·6	84·5	13·5	34·6
13·6	85·5	13·4	21·9
13·7		13·8	
13·5		13·4	
13·5		Mean value 13·595	

As the table shows, the velocity of the downward motion of the particle is always the same. On the other hand, with a given potential difference, the values of t_2 (and hence of v_2 also) vary intermittently, whilst earlier values recur later on during the up-and-down motion.

The jumps in the values of t_2 must obviously be due to *collisions* between the positively charged oil droplet and *air ions* resulting in an alteration in the charge of the moving particle. When a negative ion is taken up, the upward motion is retarded and hence the value of t_2 increased, whereas a decrease in the value of t_2 is due to the addition of a positive ion or the loss of a negative ion.

Now denoting by Q' the charge on the ion, and by v_2' the velocity of the upward motion after addition of this ion, we have in accordance with eqn. (1)

$$\frac{v_2'}{v_1} = \frac{QE + Q'E - mg}{mg}. \qquad (2)$$

Hence, on subtracting eqn. (1) from eqn. (2) we find

$$\frac{v_2' - v_2}{v_1} = \frac{Q'E}{mg}. \qquad (3)$$

In Table II the first column contains the values found for the velocity of the upward motion by dividing the interval of rise or fall, 0·5222 cm., by the values of t_2 given in Table I.[3] In the second column we find, in the first place, the differences between the successive numbers in the first column.

TABLE II

Velocity of the upward motion.	Difference.
$\frac{0·5222}{12·45} = 0·04196$	
	$\frac{0·01806}{2} = 0·00903$
$\frac{0·5222}{21·8} = 0·02390$	
	$\frac{0·00885}{1} = 0·00885$
$\frac{0·5222}{34·7} = 0·01505$	
	$\frac{0·00891}{1} = 0·00891$
$\frac{0·5222}{85·0} = 0·00614$	
	$\frac{0·00891}{1} = 0·00891$
$\frac{0·5222}{34·7} = 0·01505$	
	$\frac{0·01759}{2} = 0·00880$
$\frac{0·5222}{16·0} = 0·03264$	
	$\frac{0·01759}{2} = 0·00880$
$\frac{0·5222}{34·7} = 0·01505$	
	$\frac{0·00891}{1} = 0·00891$
$\frac{0·5222}{21·85} = 0·02396$	

Table II shows us, however, that the above differences always have one of two values, of which the one is exactly twice the other (in other experiments we find also triple and quadruple values). In the case of a given droplet the

[3] When two approximately equal values of t_2 follow successively in Table I, the mean value has been taken for Table II.

quantities m, v_1, E, and g are to be regarded as constants and hence it follows, in accordance with eqn. (3), that the *charges on the air ions* must always represent *integral multiples of an elementary charge.* Thus the data of Table II afford a *direct experimental proof of the atomistic structure of electricity.*

Moreover, the actual charge on the droplet (Q) may readily be compared with the ionic charge (Q'), and hence with the elementary quantum. We have only to start from eqn. (1), which gives the relation

$$\text{(4)} \qquad \frac{v_2 + v_1}{v_1} = \frac{Q\,E}{m\,g};$$

whence, taking eqn. (3) into account, we find

$$\text{(5)} \qquad \frac{v_2 + v_1}{v_2' - v_2} = \frac{Q}{Q'}.$$

Since the velocities v_1, v_2, and v_2' can be measured, the charge Q can therefore be compared with the smallest charge, as integral multiples of which the ionic charges manifest themselves. Millikan found that the *droplet's charge* also always proves to be an *exact integral multiple of the same elementary charge.* With the greatest accuracy he was able to detect all integral multiples of the elementary charge up to about 50, and he showed that the number of elementary quanta associated with an oil droplet could even amount to 150. As the multiplying factor increases, the accuracy of the count of course decreases, for the limits of error in this method are about a half *per cent.*, so that, with a charge of more than 200 elementary quanta, the error is already more than one electrical quantum.

Now it is evident from eqn. (1) that the absolute magnitude of the *elementary quantum of electricity* itself can be calculated, as soon as the *weight of a droplet* is known. Moreover, this weight can be determined with the aid of a relation connecting the constant velocity of fall with the radius of the particle. As already found by *Stokes* in 1845, the velocity of fall of a sphere of radius a and density ϱ in a medium of density ϱ', and with a coefficient of viscosity η, is given by the formula

$$\text{(6)} \qquad v = \frac{2}{9}\frac{g\,a^2\,(\varrho - \varrho')}{\eta}.$$

Millikan and his collaborators now arranged the experiments which served to determine the weight of a droplet in such a way that the assumptions, upon which the validity of Stokes's law is based, were satisfied with the greatest possible exactness.[4] In so far, however, as these assumptions could not be completely satisfied, the errors thereby caused were again accurately investigated. By means of the law of fall thus modified,[5] the radius of the droplets could be calculated with the greatest accuracy, and therefore, the density being known, their exact weight also followed.

Measurements extending over many years enabled Millikan in this way to determine the value of the *elementary quantum of electricity* (now generally denoted by e) with an *accuracy of one in a thousand* ; as a mean of his experimental results he found

(7) $e = 4{\cdot}774 \times 10^{-10}$ electrostatic units ($\pm\, 0{\cdot}004 \times 10^{-10}$).

It does not here seem necessary to enter any further into the details of Millikan's measurements, for in later sections various other methods will be described, which, quite independently of his researches, lead indirectly to exactly the same value for e.[6]

Historically, it may be noted that the first direct determinations of the elementary quantum of electricity were made in 1897 and 1898 by *Townsend* and *J. J. Thomson.* These researches, which gave for e an approximate value of 3×10^{-10} electrostatic units, were not concerned, however, with individual particles of matter, but with slowly falling *clouds of mist* in which *air ions* formed condensation nuclei for the drops of water. With the aid of Stokes's law of fall, the radius of the water drops could be derived from the

[4] It was particularly important that the oil droplets should be perfectly spherical—a result obtained by special artifices.

[5] For further details, *vide* Millikan's book quoted above.

[6] It also follows from this that objections which have been raised against Millikan's method can in no way call in question the existence of an elementary quantum of electricity, which has been definitely proved by numerous other methods.—Objections have been made to Millikan's work by *Ehrenhaft* and his pupils, who have experimented with spherical particles of various substances (mercury, silver, gold, oil, *etc.*) in various chemically inert gases (nitrogen, argon); the radii of the particles lay between 5×10^{-5} cm. and about 5×10^{-6} cm. (the limit of observability). From these researches, some of which were published before Millikan's first paper, Ehrenhaft concluded that as the radius of the particle (and therefore its electrostatic capacity) decreased, the charge also decreased to such an extent that it might be considerably less than the elementary quantum. On the other hand, Ehrenhaft's results have been ascribed to conceivable *variations in density*, which might be due to a spongy or flaky structure of extremely minute particles (smaller than Millikan's oil droplets) or to an adsorption of gas. *Cf.* Ehrenhaft's comprehensive papers in the *Ann. der Phys.*, **56** (1918), pp. 1–80, and **63** (1920), pp. 773–815; also Th. Sexl, *Zeitschr. f. Phys.*, **16** (1923), pp. 34–41. For the opposing point of view, *cf.* the summaries in *Die Naturwissenschaften* by R. Bär [**10** (1922), pp. 322 and 344], and by E. Regener [**11** (1923), p. 17].

velocity of fall of the mist, and hence, for a known weight of mist, the number of drops could be ascertained. Since, furthermore, the total electric charge on the cloud could be measured, the charge on an individual ion could be calculated.[7] By arbitrarily identifying this charge with an elementary quantum, a value of the right order of magnitude was obtained for the latter.

§ 2. Electrolysis, and the Mass of the Hydrogen Atom.

Long before the absolute values of atomic masses were known to physicists, chemists had been in a position to determine the *relative weights of the atoms*, *i.e.*, the *ratios* existing between their masses, with the help of the so-called *Law of Multiple Proportions*, which was established by *Dalton* in 1805. As Dalton discovered, and as he formulated in this well-known law, a definite *number*, characteristic of the element, can be allotted to every chemical element in such a way that the quantities of the elements contained in a chemical compound are related to each other as *integral multiples* of the numbers characteristic of the elements concerned.

If, following convention, we make the characteristic number for *oxygen* equal to 16 *exactly*, we then obtain the so-called *atomic weights of the elements*. The smallest atomic weight is that of hydrogen—1·008 (or, more accurately, 1·0077)—the largest that of uranium (238·2). We need not enter here further into the well-known methods by which chemists determine atomic weights.

The number by which the figures representing the atomic weights must be divided, in order to obtain the absolute masses of the atoms in grams, is nowadays generally known as the *Loschmidt number*. It is so called in honour of the physicist who, in 1865, was the first to determine approximately the absolute magnitude of atoms, by means of a theoretical method based on the gas laws (which we cannot discuss here). Hence, denoting the mass of the hydrogen atom in grams by m_H and the Loschmidt number by L, we have

$$(1) \qquad m_H = \frac{1{\cdot}0077}{L} \text{ grams.}$$

In like manner, we obtain the mass of any given atom by dividing the atomic weight of the element in question by the

[7] For further details, *vide*, *e.g.*, Millikan's book, ch. iii.

Loschmidt number. The number of grams of a substance equal to the number which represents the atomic or the molecular weight of the substance is known as a *gram-atom* or as a *mol* (gram-molecule) respectively. Thus, for example, a gram-atom of oxygen is 16 grams, a mol of water about 18 grams, and so on. The Loschmidt number therefore gives the number of atoms in the gram-atom, or of molecules in the mol of any given substance. If, now, we know the value of the elementary quantum of electricity, we are then able to calculate the Loschmidt number with great accuracy on the basis of the results of *electrochemical measurements*.

In 1833 *Faraday* derived the *fundamental law of electrochemistry* from his observations on *electrolysis*. This law states that the *amount* of a particular substance *produced* electrolytically in a given time is completely determined by the *current* and directly *proportional* to it. Furthermore, the amounts of the constituents liberated from different electrolytes by a current of given strength are "*chemically equivalent*," *i.e.*, they are related to each other as the quotients of the atomic weight and the chemical valency. Thus a current which liberates 1·008 grams of univalent hydrogen in a unit of time will liberate 8 grams of bivalent oxygen of atomic weight 16 in the same interval, *etc.*

The fundamental law of electrochemistry can be very simply explained on the assumption that, in an electrolyte, the molecules are entirely or partially split up, or *dissociated*, into oppositely charged components—the so-called *ions*—and that the conduction current in the electrolyte consists in a *transport* of such ions. The positive ions thus wander in the direction of the current to where it leaves the liquid, at the so-called *cathode*, whereas the negative ions wander in the opposite direction to where the current enters the liquid, at the so-called *anode*. Faraday's Law then follows at once on the hypothesis that the charge on all univalent ions is one elementary quantum of electricity, that on bivalent ions two, and that on trivalent ions three elementary quanta, *etc.*

Denoting the number of atoms liberated in a unit of time at the cathode by N and the valency by z, the current, which is the quantity of electricity passing the cathode in unit time, is given by

$$I = N z e. \tag{2}$$

On the other hand, the mass of a single atom is given by the quotient of the atomic weight and the Loschmidt number, *i.e.*, by the quantity A/L, where A is the atomic weight. The amount of the substance in question liberated in a unit of time is therefore determined by the relation

$$M = \frac{NA}{L}, \tag{3}$$

or, in view of eqn. (2), by

$$M = \frac{A}{z}\frac{I}{Le}. \tag{4}$$

Now this equation is an actual expression of the fundamental law of electrochemistry. For, since L and e are universal constants, it follows on the one hand that, with a given substance, and therefore with given values of A and z, the mass (M) liberated in unit time is proportional to the current (I); and on the other hand it follows that, with a given current, the value of M is proportional to the quotient of atomic weight and valency.

Since, in accordance with what we have said above, the quantity M/A represents the number of gram-atoms, we also recognize from eqn. (4) that the *product of the Loschmidt number and the elementary quantum of electricity* simply represents the *current*, measured in electrostatic units, which *liberates one gram-atom of a univalent element* in a unit of time.

Very accurate measurements, in which principally silver has been used, have demonstrated that a gram-atom of a univalent substance would be deposited in a unit of time by a current of 96,494 ampères. In order to express this current in electrostatic measure, we have to multiply it by one-tenth of the velocity of light. Since the velocity of light is equal to $2{\cdot}999 \times 10^{10}$ cm. sec.$^{-1}$, we thus find

$$L\,e = 2{\cdot}894 \times 10^{14} \text{ E.S.U.} \tag{5}$$

Hence, introducing the value already given for the elementary quantum of electricity, we obtain for the *Loschmidt number*

$$L = 6{\cdot}062 \times 10^{23}, \tag{6}$$

with an error which can scarcely be more than 1 in 1,000.

It follows from eqn. (1) that the *mass of the hydrogen atom* is

$$m_{\text{H}} = 1{\cdot}662 \times 10^{-24} \text{ grams.} \tag{7}$$

For the ratio of charge to mass, *i.e.*, for the so-called *specific charge of the ionized hydrogen atom*, we have, in accordance with eqn. (1),

$$(8) \qquad \frac{e}{m_H} = \frac{Le}{1 \cdot 0077},$$

or, by eqn. (5),

$$(9) \qquad \frac{e}{m_H} = 2 \cdot 871 \times 10^{14} \text{ E.S.U./gram.}$$

§ 3. Negative Corpuscular Radiation and the Electron.

In 1858 *Plücker* discovered the remarkable rays which are emitted by the cathode in highly evacuated Geissler tubes, and which are therefore known in general as *cathode rays*. Their properties were more closely investigated, particularly by *Hittorf* (1869) and *Crookes* (1879), both of whom made the surprising discovery that these rays, which ordinarily travel in *straight lines*, are very easily *deflected* by a *magnet*. Moreover, *Goldstein* (1876) showed that they may also be *deflected* by an *electric field*.[1]

As was first pointed out by *J. J. Thomson* and *Kaufmann* (1897), all these properties of the cathode rays can be simply explained on the assumption that they consist of quickly moving *electrically charged particles*, which at the same time possess *inertial mass*. Their rectilinear propagation follows from their inertia, and the observed deflections both in a magnetic and in an electric field necessarily follow from their electrical charge; the sense of the deflection in both cases shows that the *charge* must be *negative*. Unlike rays of light or heat, which are forms of electromagnetic waves, the cathode rays therefore represent a so-called *corpuscular radiation* consisting of a stream of inertial particles.

Now it follows from the theory of electricity, without the aid of any kind of atomistic hypothesis, that, in a *transverse homogeneous magnetic field*, a moving electrically charged mass must describe a *circular path* with constant velocity in a plane at right angles to the magnetic lines of force. The radius of the circle is given by the simple relation

$$(1) \qquad r = \frac{cv}{\gamma H},$$

where H is the magnitude of the magnetic strength of

[1] The study of the cathode rays was greatly aided by an artifice of *Lenard's*, which consisted in letting the rays escape from the tube through a piece of thin metal foil inserted in the wall of the tube.

field, γ the specific charge of the moving mass, and v its velocity.

It follows likewise from the theory of electricity, irrespective of any kind of atomistic considerations, that, in a *transverse homogeneous electric field*, a moving charged body describes a *parabola*, the curvature of which depends only on the specific charge and the velocity, in addition to the electric field-strength. If we take the original direction of the path as x-axis, the direction of the electric force as z-axis, and the position of the moving body at the time $t = 0$ as origin, we have then the relation

$$\frac{z}{x^2} = \frac{1}{2}\frac{\gamma}{v^2}E, \tag{2}$$

where E is the magnitude of the electric field-strength.[2]

As a matter of fact, observations show that the particles in the cathode rays do move in a circular path in a transverse magnetic field, and in a parabola in a transverse electric field. Hence, if we apply the two equations to cathode rays, and regard γ and v as unknown, we may calculate the velocity and the specific charge of the particles from the known measured values of the magnetic and electric strengths of field.

Exceedingly large values are obtained for the *velocity* of the cathode rays, extending from a few *per cent.* to about *one-half of the velocity of light.* Naturally, the velocity depends on the potential applied to the cathode tube (*viz.*, on the potential difference through which the particles travel). If V be the tube potential and Q the charge on a cathode particle, the product $Q\ V$ must, of course, be equal to the kinetic energy of the particle as it leaves the tube, *i.e.*, to $\frac{1}{2}\,mv^2$. Hence

$$v = \sqrt{2\gamma\ V}. \tag{3}$$

Thus the velocity varies as the square root of the tube potential.

The value of the *specific charge*, as obtained from measurements according to eqns. (1) and (2), is

$$\gamma = 5{\cdot}3 \times 10^{17}\ \text{E.S.U./gram.} \tag{4}$$

On comparing this value with that obtained for the specific charge of the ionized hydrogen atom, we see that it is about 1,800 times as large. If we make the very likely assumption that the *charge on a cathode ray particle* amounts to *one*

[2] An exact derivation of eqns. (1) and (2) may be found, *e.g.*, in the author's "*Introduction to Theoretical Physics,*" Vol. I., § 71.

elementary quantum of electricity, we must then ascribe to the particle a *mass* 1/1800th of that of the *hydrogen atom.*

When this assumption was first suggested, it appeared to physicists an extremely bold one, because, up to about the end of the 19th century, the mass of the hydrogen atom had been regarded as the smallest mass possible. In 1896, however, the discovery of the *Zeeman effect* (which we shall discuss later), which consists in the resolution of spectral lines in a magnetic field, led to the same conclusion. Moreover, the subsequent progress of theoretical physics has appeared entirely to confirm all deductions made on the hypothesis of smallest particles with a mass 1/1800th of that of the hydrogen atom. These particles with a negative charge of one elementary quantum of electricity are now generally known as *negative electrons,* or simply as *electrons.*[3] It follows from eqn. (4) and the known value of the elementary quantum of electricity that the *mass* of an electron, which is generally denoted by m in atomic theory, is given by

$$m = 9{\cdot}0 \times 10^{-28} \text{ grams.} \tag{5}$$

As was mentioned above, the velocity of the cathode rays depends on the tube potential ; and since the latter cannot be indefinitely increased, an upper limit is thus set for the velocity of the rays, this limit depending on the state of experimental technique. Besides the artificially produced negative corpuscular rays, however, physics is also acquainted with spontaneously occurring rays, likewise consisting of electrons, but possessing a much higher velocity. These rays form part of the radiation emitted by so-called *radioactive substances.*

In 1896 *Becquerel* made the surprising discovery that *uranium ores* constantly emit rays without the intervention of any outside agency. These rays are able to blacken a photographic plate through an opaque covering, and also to ionize the air, thus rendering it conducting, so that a charged electroscope set up close at hand is discharged. Two years later (1898), *Pierre* and *Marya Curie* found that the so-called Becquerel rays (as they were originally termed) are, in point of fact, principally emitted by a metal contained in the pitchblende. At that time this metal was still unknown, but, after tedious researches, M. and Mme. Curie isolated it from pitchblende and named it *radium.*[4] A closer investigation

[3] The term " electron " is due to Stoney (1881).

[4] *Metallic* radium was not obtained until 1910, by Mme. Curie ; up till then only salts of radium were known.

of the " radioactive radiation " emitted by radium and other substances showed that it consists of *three* entirely different kinds of rays, which are distinguished as *alpha-*, *beta-*, and *gamma-rays ;* a separation of the three kinds of rays is rendered possible by their *different behaviour in a magnetic field.* Whereas the γ-rays are not deflected at all, and are thus to be regarded as a radiation of electromagnetic waves, the α- and the β-rays are found to be corpuscular rays that are deflected by a magnet. Furthermore, it appears that the α-rays (which will be discussed later) carry positive charges, whereas the β-rays carry negative charges.

The *velocity* and *specific charge* of the *beta-rays* could be determined by the same method as in the case of cathode rays. The value obtained for the specific charge was the same as for cathode rays, which apparently proved the β-rays to consist likewise of *electrons.* They are thus essentially identical with cathode rays, but they are distinguished from them by a *greater velocity,* for which values ranging from 30 *to* 99·8% *of the velocity of light* have been obtained.

The motion of β-particles is the *most rapid material motion* known in physics, and for this enormous velocity so closely approaching that of light [5] are clearly revealed the *deviations from classical mechanics* which, according to the *theory of relativity,* are to be expected in the case of such rapid motions. It follows from this theory that *mass* depends on *velocity* in accordance with the relation

$$m = \frac{m_0}{\sqrt{1 - \frac{v^2}{c^2}}}, \tag{6}$$

where m_0 denotes the value of the mass for velocities small compared with the velocity of light (c). This formula shows that, with a velocity of $v = c/2$, the mass is some 1·15 times as large as m_0, whilst, with a velocity of 99·8% of the velocity of light, the mass is as much as 16 times as large. Measurements by *Kaufmann* and other investigators have actually established the increase in mass with increasing velocity, as demanded by the relativity theory, and have also confirmed eqn. (6) quantitatively.[6]

Essentially the same as cathode and β-rays are the negative corpuscular rays which are a consequence of the so-called

[5] It appears from the theory of relativity that velocities greater than the velocity of light are impossible, as is also shown by eqn. (6).

[6] An increase in mass connotes a decrease in γ.

photo-electric effect. In continuation of an earlier research by *Hertz*, *Hallwachs* in 1888 made the important observation that a *metal plate* becomes *positively charged* when illuminated with *ultra-violet light*. *Lenard* recognized the cause of this phenomenon to be a negative corpuscular radiation emitted by the illuminated metal. This radiation was found to have the same specific charge as cathode rays, so that obviously it must also consist of *electrons*. Their velocity, however, proved to be much smaller than that of cathode rays, only amounting to some thousandths of the velocity of light.

Every corpuscular radiation of definite velocity and definite specific charge may be characterized by a definite value of the electric potential, which may be called the *equivalent potential* and, when multiplied by the charge of the radiated particle, gives its kinetic energy. Thus, in accordance with eqn. (3), we have the relation

$$V = \frac{v^2}{2\gamma}, \tag{7}$$

or, when the potential is expressed in volts,

$$V = 150\frac{v^2}{\gamma} \text{ (V in volts, γ in electrostatic measure).} \tag{8}$$

In the case of cathode rays we find, of course, as equivalent potential the tube potential, or about 30,000 volts for velocities amounting to one-third of the velocity of light. The potential corresponding to the fastest β-rays would be about 4,000,000 volts, for v is then almost equal to c, and γ is 16 times smaller than in the case of cathode rays (in consequence of the increase in mass). On the other hand, the rays produced by ultra-violet light have an equivalent potential of only a few volts.

§ 4. Positive Corpuscular Radiation.

The negative rays which we have hitherto been discussing find their counterpart in the *positive corpuscular radiation*. This was discovered by *Goldstein* in 1886, in the form of the so-called *canal rays*. Like the cathode rays, the canal rays emanate from the cathode in a highly evacuated Geissler tube, but travel in the opposite direction to the cathode rays.[1] It was, moreover, later observed that the anode itself emits positive rays, known as *anode rays*.[2]

[1] In order that the rays may escape from the cathode, the latter must be perforated by small canals; hence the name "canal rays."

[2] The anode rays were discovered in 1906 by *Gehrcke* and *Reichenheim*.

There is, however, otherwise no essential difference between canal rays and anode rays. Their velocity and specific charge can be determined by the same method as in the case of negative corpuscular rays, the sense of deflection clearly demonstrating that the particles in canal and anode rays are *positively charged*. Their *velocity* is found to be a few thousandths of the velocity of light.

Experiment shows the value of the *specific charge* to be much smaller than in the case of the negative rays; furthermore, this value depends on the particular gas of which traces remain in the evacuated tube. The largest value of the specific charge is obtained when the tube is filled with hydrogen before evacuation; in that case the value is identical with that obtained from electrolytic measurements for the specific charge of the hydrogen ion. If the tube be filled with mercury vapour before evacuation, the specific charge is found to have a value 1/200th of the above, corresponding to the fact that the atomic weight of mercury is 200. The particles in positive corpuscular rays must therefore be regarded as *electrically charged atoms*, the charge on each amounting to one, or sometimes to several elementary quanta.[3]

Now, in transverse electric and magnetic fields the particles in a beam of corpuscular rays must be deflected, and, if the electric field be arranged at right angles to the magnetic field, the deflections will be such that all particles with the *same specific charge* will strike a plate, placed at right angles to the beam, along a *parabolic curve*. Hence, if a positive radiation be composed of atoms with different specific charges, these must form parabolas of different curvatures, and upon this fact rests the possibility of *positive ray analysis*, as was first recognized and applied by *J. J. Thomson*.

The method of positive ray analysis has been developed by *Aston* (1919) into one of the most important methods of experimental research.[4] Owing to the focussing effect

[3] In addition to charged atoms, we may also meet with charged atomic complexes, *e.g.*, molecules of hydrochloric acid which possess a positive charge of one elementary quantum, and so on.

[4] In Aston's method the electric and magnetic fields are applied so that the deflections are at 180° instead of at 90° to each other. A very thin ribbon of rays is spread out by the electric field into an "electric spectrum," and a group of the deflected particles is allowed to pass through the magnetic field, which is so arranged that the resulting deflection is opposite in direction to that produced by the electric field. It may be shown that all particles with the same specific charge come to a real focus, and that the positions of the foci so generated lie along a straight line, which is sharply defined on a suitably placed photographic plate.

obtained, it is possible to produce much greater resolution with his form of apparatus than with the original method. In this way the masses of arbitrary particles can be exactly compared with each other; and in addition to some other very important experimental results which we shall discuss fully later, Aston's researches established the most important fact that the *masses of arbitrary positive ray particles* are always integral multiples of the mass of the hydrogen atom (or, more accurately speaking, *almost always whole numbers in terms of the mass of the oxygen atom, taken as* 16).

Like cathode rays, canal and anode rays are artificial products of experimental physics. A *natural* counterpart to the negative β-rays is represented by the *positive alpha-rays* which, as we have already mentioned, are emitted similarly to the β-rays by radioactive substances. Their positive electrical nature is proved by the fact that they are deflected by a magnet, but in the opposite sense to β-rays.

The *velocity* of α-rays generally lies between *five* and *seven per cent. of the velocity of light*, varying with the chemical nature of the radioactive substance emitting them. Measurements of the *specific charge* yield a value of

$$(1) \qquad \gamma = 1{\cdot}44 \times 10^{14} \text{ E.S.U./gram.}$$

Thus the specific charge of the α-particle is *half as large as that of the ionized hydrogen atom.*

Now it is a fact of great importance that the α-particles emitted by a radioactive substance can be *counted*, for the *charge on an individual α-particle* can then be directly determined. This is rendered possible by a remarkable phenomenon known as *scintillation*. If we bring a radioactive preparation into the neighbourhood of a fluorescent screen (of zinc blende or a very thin section of diamond), we observe a continual flashing of discrete points of light. This suggests that each flash of light is due to the impact of a *single* α-particle.[5] We can place the radioactive preparation behind a screen of some substance which absorbs the α-radiation, and in which there is a small hole allowing an exactly calculable fraction of the total α-radiation emitted by the preparation to pass through. This fraction is allowed to strike a small portion of the scintillation screen, which is observed under a microscope. In this way we can readily

[5] This assumption is justified by the fact that the scintillation method gives the same value for the disintegration constant of radium as do other methods (which we shall discuss subsequently).

count the number of scintillations occurring in a definite time, and, knowing the size of the hole, we can calculate how many α-particles are emitted every second by a given preparation. Determinations of this kind were first made by *Rutherford* and *Geiger* in 1908.

Furthermore, by cutting out the β-rays emitted simultaneously with the α-rays by means of a magnetic field, a metal plate can be charged positively by the radioactive radiation. In order to find the electric charge on a single α-particle, we have then only to divide the total charge conveyed to the plate in a given time by the known number of α-particles which strike the plate in that time. Rutherford and Geiger's measurements showed that the value of the charge on an α-particle is exactly *double* that of the *elementary quantum of electricity*. Since, on the other hand, the specific charge is half that of the ionized hydrogen atom, we must ascribe to the α-particle a *mass equal to that of four hydrogen atoms*, and a *positive charge of two elementary quanta*.

On *passing through ordinary substances*, *α-rays* are far more strongly *absorbed* than β-rays, and, moreover, the nature of the absorption is entirely different. For, whereas β-rays are absorbed by matter in accordance with an *exponential law*, the action of α-rays (such as the blackening of a photographic plate or the ionization of air) ceases quite suddenly, as soon as the rays have penetrated a certain distance into a given medium. This distance is known as the *range*, and it depends on the chemical character of the radiating substance as well as upon the nature of the absorbing medium. The greater the initial velocity of the α-rays on their emission from the radioactive substance, the larger is the range, which, as *Geiger* has empirically discovered, is *proportional to the cube of the initial velocity*. In ordinary air the ranges of known α-rays lie between 2·7 and 11·3 cm. ; in hydrogen the α-rays of RaC have a range of 28 cm. (at room temperature and ordinary pressure). An aluminium plate 0·06 mm. thick suffices to absorb completely all but the fastest α-rays, whereas β-rays can penetrate through aluminium plates some millimetres in thickness.

The passage of α-rays through matter has been investigated with particular success by *C. T. R. Wilson* (1912). His experiments were based upon the well-known fact that, in *supersaturated water vapour*, electrically charged particles serve as *condensation nuclei* for *water drops*. Thus, on bringing a radioactive preparation into a chamber containing supersaturated water vapour, Wilson was able to *photograph* the collection of water drops created by the passage of an α-particle. A photograph of the streaks of mist thus produced gives also a picture of the *actual paths of the α-particles*. In general the photographs (Figs. 1 and 2) show *straight lines*, many of which, however, appear *bent at the end* through fairly large angles. This fact, which we shall discuss in detail later, is of fundamental importance for modern atomic theory.

Besides canal, anode, and α-rays, a fourth kind of positive

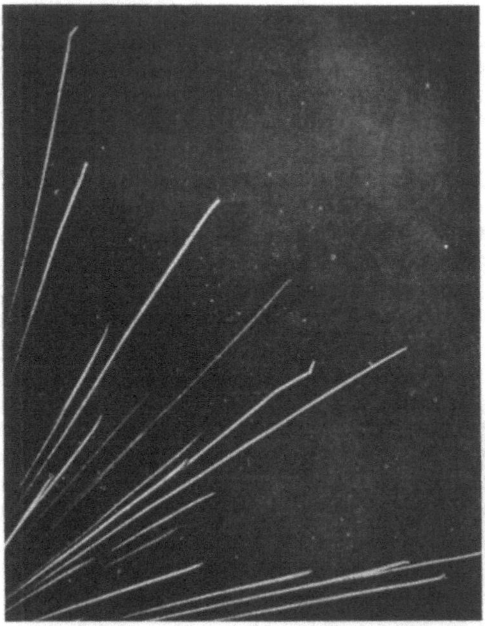

FIG. 1.

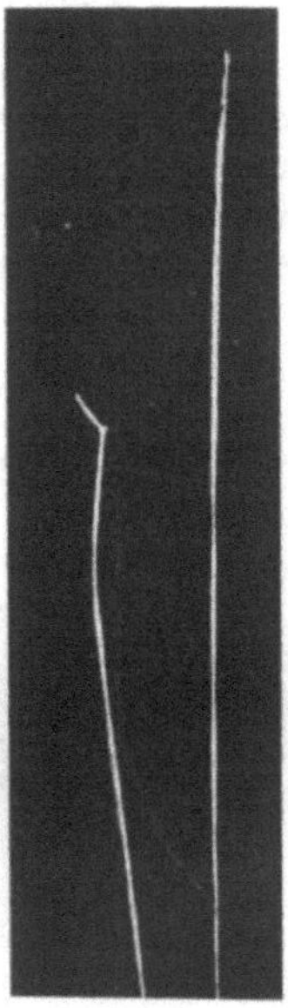

FIG. 2.

Paths of α-rays.

corpuscular rays is known in physics, *viz.*, the so-called *H-rays*, discovered by *Marsden* in 1914. When he passed the α-particles emitted by radium C through *hydrogen*, in which their range is 28 cm., Marsden was able to detect scintillations on a screen even at a distance of 80 cm. These scintillations could obviously not be due to the α-rays, but must have been caused by other rays which, in their turn, were produced by the α-particles. A determination of the specific charge gave the same value for these newly discovered rays as is obtained from electrolytic measurements for hydrogen ions. From the chemical symbol for hydrogen, these rays are therefore called H-rays.

Equivalent potentials can easily be calculated for positive corpuscular radiation. From eqn. (1) and § 3 eqn. (8) we find for α-rays, whose velocity, as already mentioned, amounts to 5 to 7% of the velocity of light, a value of about two to four million volts. If we wish to express in ergs the kinetic energy of a rapid α-particle possessing a velocity of 2×10^9 cm./sec., we have to multiply half its mass (*i.e.*, twice the mass of a hydrogen atom) by the square of its velocity. We thus find

(2) $$E_{kin} = 1{\cdot}3 \times 10^{-5} \text{ ergs.}$$

The velocity of the fastest β-particles is, it is true, some 15 times as large, and therefore the square of the velocity about 200 times as large. On the other hand, their mass is some 7,000 times as small, and even when we take into account the 16-fold increase in mass for the enormous velocity of 99·8% of the velocity of light, the mass is still some hundreds of times as small as that of an α-particle. The kinetic energy of β-particles is therefore considerably less than that of α-particles.

§ 5. The Elementary Quantum of Action.

During an investigation of the *photo-electric effect* in 1902, *Lenard* had already observed the remarkable fact that the *velocity* of the electrons liberated by ultra-violet light is entirely *independent of the intensity of the light used.* The velocity cannot be increased by raising the intensity, but only by raising the frequency of the light. Increasing the intensity augments only the *number* of liberated electrons, but not their velocity.

Quantitative measurements of the photo-electric effect are based on the fact that the emission of the electrons from the metal plate subjected to the radiation can be prevented by giving the plate a *positive charge.* With a given frequency of the incident light, we can thus determine for a plate of a given metal a quite definite *counterpotential,* on reaching which the emission of electrons just comes to a complete stop. If V be the counterpotential, experimental observation shows that *V increases according to an exact linear relation with the frequency of the light.* We may write

(1) $$V = C\nu - P,$$

where C and P are positive constants. We find from experiment that only the constant P depends on the nature of the metal,[1] whereas the constant C is independent thereof and must therefore represent a *universal constant*. Now the product $e\,V$ denotes the kinetic energy of the most rapid electrons liberated, which suggests a multiplication of eqn. (1) by e. If we write h for the product $C\,e$, eqn. (1) becomes

$$(2) \qquad e\,(V + P) = h\,\nu.$$

Since the product $e\,V$ represents work or energy, whilst frequency has the reciprocal dimensions of time, the dimensions of the constant h must be those of a *product of energy and time*. Such a product in mechanics is termed *action*, and the constant h is therefore called the *elementary quantum of action*. The constant h thus forms a *fourth* elementary quantum, in addition to the elementary quantum of electricity and the masses of the electron and the hydrogen atom. Its value is found from measurements of the photo-electric effect to be

$$(3) \qquad h = 6{\cdot}55 \times 10^{-27} \text{ erg sec.}$$

The hypothesis of the elementary quantum of action was introduced into theoretical physics by *Planck* in 1900. By means of this hypothesis Planck succeeded in solving a difficult fundamental problem in the theory of thermal radiation, which, however, we cannot discuss here. For this purpose he assumed that the emission of wave radiation occurs *discontinuously* in such a way that *elements of energy* of magnitude $h\,\nu$ play a part in the process.

In 1905, *Einstein* extended Planck's conception by introducing the idea of *light-quanta* whose energy is likewise to be determined by the product $h\,\nu$, where ν is the frequency of the light.[2] Einstein formulated the *law* that every *mutual transformation between wave radiation and corpuscular radiation* takes place in such a way that either a *single light-quantum* is utilized in the liberation of a single electron, or, in the converse case, a single light-quantum is produced by the energy of a single electron.

The first case occurs in the photo-electric effect, where wave radiation is transformed into corpuscular radiation, and, as a matter of fact, later exact measurements by

[1] It is true that the presence of adsorbed or occluded gases has a great influence.

[2] Planck had already been able to determine the value of h from his theory of thermal radiation.

Millikan have demonstrated the accurate fulfilment of Einstein's law for the photo-electric effect. For, in the empirically obtained eqn. (2), a light-quantum of the exciting light has been equated to the sum of the maximum kinetic energy ($e\ V$) with which a liberated electron leaves the metal subjected to radiation, and the work ($e\,P$) which the electron must perform in escaping from the metal. The *constant* P, which likewise represents a potential, when multiplied by the elementary quantum of electricity is thus a measure of the *work expended in freeing an electron*, and it thus becomes intelligible why the value of P should agree in order of magnitude with the difference of potential between two metals not too close together in Volta's well-known potential series. The values of P, which is termed the "*contact potential*," amount to only a few volts for the various metals. The smallest value is for *potassium*, in which, as is well known, the photo-electric effect is the greatest. In accordance with eqn. (2), the expression $e\,P/h$ determines the *limiting frequency* below which the metal cannot be photo-electrically excited.[3]

We shall not discuss until later the converse phenomenon to the photo-electric effect, the transformation of corpuscular radiation into wave radiation; this phenomenon also confirms Einstein's law of light-quanta. Moreover, Einstein's hypothesis also affords a simple explanation of a relation holding for the transformation of primary into *secondary wave radiation.* According to Einstein's law, *one secondary light-quantum* must arise *from each primary quantum.* If this be the case, however, the *principle of energy* teaches that the *secondary frequency cannot be greater than the primary;* it must be *smaller* when energy is also expended for other purposes during the transformation. Secondary wave radiation is termed *fluorescent light.* As a matter of fact, *Stokes* in 1852 established the *rule*, known by his name, that the *wave-length* of the fluorescent light is *longer* than that of the exciting light.[4]

[3] This limit has been determined for many substances. In the case of copper, for example, it lies at about 3,000 Å.U.; in the case of sodium, at about 6,000 Å.U. The limiting wave-length for potassium is at any rate greater than 10,000 Å.U.

[4] For apparent exceptions, *cf.* § 36.

Chapter II

THEORY OF THE HYDROGEN ATOM

§ 6. The Nuclear Atom.

As we have seen, the phenomena of corpuscular radiation render manifest the existence, on the one hand, of particles with a *negative charge* of one elementary quantum of electricity and with a mass invariably equal to 1/1,800th of that of the hydrogen atom, and, on the other hand, of corpuscles with a *positive charge* of one or of a few elementary quanta and possessing a *mass* which is always approximately an *integral multiple* of that of the hydrogen atom.

It would appear that *atoms* must be *composed* of these negative and positive corpuscles, and important clues as to the nature of this structure are afforded by the previously described observations on the *passage of alpha-rays through matter*. Now these observations show that an α-particle can describe *rectilinear paths* of a few centimetres in air. A cubic centimetre of air, however, contains approximately 5×10^{19} atoms, as follows from the known density and from the known value of the Loschmidt number.[1] On the other hand, it has long been known from the kinetic theory of gases that the diameter of an atom must be of the order of magnitude of 10^{-8} cm., so that, in air, about 1/100,000th part of the total volume is occupied by the atoms, and a line 1 cm. long therefore passes through thousands of atoms. Hence an α-particle is capable of *penetrating through many thousands of atoms* without its direction being appreciably affected. Occasionally, however, as Wilson's photographs show, *a single atom deflects the α-particle through a very large angle.*

This remarkable fact can be readily explained, on the basis of *Coulomb's Law*, by the assumption that the deflection is caused by an *electrical force* exerted on the electrically positive α-particle by an *atomic nucleus*, the *volume* of which constitutes *only a very small part of the total volume of the*

[1] The number of molecules is half as large.

atom. We might, at first, equally well suppose the electrical force to be one of attraction as of repulsion, and hence the atomic nucleus to be either negative or positive.

Now it follows from the principle of action and reaction that the extraordinarily large deflections of the α-rays actually observed are only possible if the *mass of the nucleus* is considerably *larger* than the mass of the deflected α-particle. We must therefore assume that the *mass of the atom* is principally *concentrated in the nucleus*.

Again, as the mass of the electrons is vanishingly small in comparison with that of the positive electrical particles, we must suppose that the nuclei of the atoms are electrically *positive*. Since, in general, the atoms as a whole are electrically *neutral*, this positive charge must be *compensated* by negative charges; and, because negative particles are found *only* as electrons in cathode and β-rays, the positive nuclear charge must obviously be compensated by *electrons* surrounding the nucleus. Provided that the *motion* of these electrons *around the nucleus* be sufficiently rapid, the electrical force of attraction which would otherwise cause them to unite with the nucleus will be counterbalanced by the *centrifugal force*.

These considerations lead to a *model of the atom* which was first introduced into theoretical physics by *Rutherford* in 1911. According to this model, the atoms consist of positively charged nuclei around which, to borrow a frequently used simile, electrons revolve like planets around the sun.[2]

The *simplest* imaginable atomic model is obviously one in which a *single electron* revolves about an *atomic nucleus* charged with *one positive elementary quantum*. The idea at once suggests itself to associate this model with the *lightest* of all atoms, the *hydrogen atom*, and it is actually found that all conclusions which may be theoretically derived from this assumption are completely confirmed by experiment. Hence, we may also speak of *hydrogen nuclei*, instead of *positive*

[2] Rutherford based his atomic model, which he had developed a year before Wilson's researches, on his own observations on the passage of α-rays through metal foils.—To-day only a historical significance is attached to *J. J. Thomson's model of the atom* which, at that time, was a subject of much discussion. Thomson supposed the atom to consist of a uniformly dense spherical volume charge of positive electricity, within which electrons described circular orbits about the centre of the sphere. The force acting on an electron is then directly proportional to its distance from the centre (because only a portion of the spherical charge is effective). Although the assumption of such forces proved most fruitful for the explanation of dispersion and magneto-optical phenomena, Thomson's atomic model had to be relinquished when observations on the passage of α-rays through matter rendered it untenable.

corpuscles or electrons, but the word "*protons*" is now generally adopted.[3]

§ 7. Bohr's Model of the Hydrogen Atom.

In Rutherford's model of the hydrogen atom just described, there at first appeared to be really no way of determining the radius of the atom, or, more accurately, the radius of the orbit described by the electron.[1] In 1913 *Bohr* succeeded in solving this riddle, by *applying the hypothesis of the elementary quantum of action to Rutherford's theory of the atom.*

Bohr's fundamental assumption states that the *angular momentum of the normal hydrogen atom, multiplied by* 2π, *shall be equal to the elementary quantum of action;* this relation determines the normal radius of the hydrogen atom. Moreover, in accordance with an elementary relation of mechanics, the angular momentum is given by the product of the mass of the electron (m), its orbital velocity (v), and the radius of its orbit (a).

In addition to the normal state, however, *other quite definite states* are considered possible, *viz.*, those for which 2π times the *angular momentum* is *equal to an integral multiple of the elementary quantum of action.* These *abnormal* states are described as *multiple-quantum* states, and the corresponding whole number is called the *quantum number* of the state. Thus we speak of a two-quantum, a three-quantum, a four-quantum state, and so on; in the hydrogen atom the *fundamental* or *normal orbit* of the electron is a *one-quantum* orbit.

We will now calculate the values of the *radius of the orbit* and of the *energy* for a state of arbitrary quantum number in Bohr's atomic model. In view of later applications we shall consider, besides the hydrogen atom itself, the more general case in which the nucleus, around which a *single* electron is revolving, has a positive charge of z elementary electrical quanta; such an atom may be called a *hydrogen-analogue.* Since z is greater than unity, it obviously cannot be neutral, but must behave as a *positive ion.*

The forces acting on the revolving electron are, on the one hand, the *electrical attraction of the nucleus,* and, on the other hand, the *centrifugal force.* In accordance with Newton's Third Law of Motion, the *two forces* must be *equal.* By

[3] Rutherford suggested the term "*protons*" for positive electrons This terminology renders the use of the single word "electron" preferable to the term "negative electron" for the negative constituent.

[1] In Thomson's model of the atom, which was based on a sphere of given radius, this difficulty, of course, did not arise.

Coulomb's Law, the force of attraction is equal to $z\,e^2/a^2$; the centrifugal force is $m\,v^2/a$, where m is the mass of the electron.

Equating the two expressions, we find

$$(1) \qquad m\,v^2\,a = z\,e^2 .$$

Again, in a permissible orbit of the quantum theory, the following relation must be obeyed, since it is postulated that the angular momentum, multiplied by 2π, shall be an integral multiple of the elementary quantum of action :—

$$(2) \qquad m\,a\,v = \frac{n\,h}{2\pi}$$

If the *quantum number* (n) be given, eqns. (1) and (2) contain as unknowns only the orbital radius (a) and the linear velocity (v), so that these two magnitudes can be determined from the two equations. We may distinguish by the subscript n the values assumed by these magnitudes for the n-quantum state. The value of a_n is obtained by squaring eqn. (2) and dividing by eqn. (1); we thus find

$$(3) \qquad a_n = \frac{n^2\,h^2}{4\pi^2\,z\,e^2\,m} .$$

The *radii* of the permissible orbits therefore vary as the *squares of the quantum numbers.*

We find the *velocity* in the n-quantum orbit on dividing eqn. (1) by eqn. (2); we have then

$$(4) \qquad v_n = \frac{2\pi\,z\,e^2}{n\,h} .$$

The linear velocities therefore vary inversely as the quantum numbers. In the case of the normal hydrogen atom, for which both z and n are unity, we find, on introducing the known values of e, h, and m into eqns. (3) and (4), that the radius of the orbit is about 5×10^{-9} cm., and the velocity about 2×10^8 cm./sec., or somewhat less than one-hundredth of the velocity of light.

Finally, let us calculate the *energy* possessed by the revolving electron in the n-quantum orbit. This energy is made up of kinetic and potential energy. The *kinetic* energy is found by multiplying half the square of the velocity by the mass (m). The *potential* energy follows from Coulomb's Law :—

$$(5) \qquad E_{pot} = -\frac{z\,e^2}{a} .$$

A *negative sign* has to be prefixed because, on account of the attraction between nucleus and electron, an *increase in the distance* a requires an *addition of energy*, and the energy can only increase with a (to which it is inversely proportional) if it be *opposite in sign.* In spite of the negative sign, the energy itself is, of course, not negative, for in eqn. (5) there has still to be added a positive constant which has the same value for all states. Besides, the potential energy represents the *work* which would be required to remove the electron, were it at rest, to an infinite distance.

On comparing eqns. (1) and (5) we have

(6) $$E_{pot} = -m v^2 .$$

The potential energy is therefore twice as large as the kinetic energy, and opposite in sign ; hence the *total energy* (the sum of kinetic and potential) is given by

(7) $$E_n = -\frac{m v^2_n}{2},$$

or, according to eqn. (4), by

(8) $$E_n = -\frac{2\pi^2 z^2 e^4 m}{n^2 h^2} .$$

The *energy values* belonging to the different quantum orbits thus vary *inversely as the squares of the quantum numbers.*

§ 8. The Spectral Formula of Bohr.

In Bohr's model of the atom, every *transition between two orbits with different quantum numbers* is associated with a *change in the energy of the atom.* On account of the negative sign in § 7 eqn. (8), larger quantum numbers correspond to larger values of the energy. Hence, if the *quantum number be lowered* in the transition, *energy* is *liberated.* On the other hand, only an *addition of energy* will permit of a transition in which the *quantum number is raised.*

Now it is a fundamental assumption in Bohr's theory that the liberation of energy by a transition of this kind must always produce a *single light-quantum,* and that, in the reverse process, the addition of energy will always be effected by the taking up of a single light-quantum. If, therefore, we consider two orbits whose quantum numbers are n and s, where $n > s$, radiation of a definite frequency (ν) will be *emitted* for a transition from the n-quantum to the s-quantum orbit, whilst, in the reverse transition, radiation of the same frequency ν will be *absorbed.* The *frequency,* however, is determined by the so-called *frequency condition of Bohr* just quoted, in that, when the frequency is *multiplied by the elementary quantum of action*, the product is *equal to the difference in the energy values*. We thus have

$$(1) \qquad h\nu = E_n - E_s.$$

On introducing the values given for the energy in § 7 eqn. (8), the frequency condition assumes the form of the *spectral formula* derived by Bohr in 1913 :—

$$(2) \qquad \nu = \frac{2\pi^2 z^2 e^4 m}{h^3}\left(\frac{1}{s^2} - \frac{1}{n^2}\right).$$

Since both s and n can run through the whole series of integers, there results a *double manifold of frequency values*. This double manifold is defined as the *line spectrum* of the atom. It may occur either as an *emission spectrum* or as an *absorption spectrum*. It is composed of *series* which, in their turn, represent simple manifolds and are obtained by giving the one quantum number (s) a fixed value, whilst the other quantum number (n) passes through the series of whole numbers.

Bohr's frequency condition leads at once to the *representation of spectral lines by spectral terms*. For, by eqn. (1), the frequency of every line can be written as the *difference between two terms*, each of which represents one of the possible energy values divided by the elementary quantum of action. The whole of the lines in the spectrum can thus be *referred to a simple manifold of terms*. Each series can hence be represented as a difference between a *fixed term* and a *variable term*, the *variable number* of which passes through the series of integers. The fixed term thus determines the *limiting frequency of the series*, which is approached by the lines as the variable number increases.

If, now, n, r, and s be three arbitrary whole numbers, the identity

$$(3) \qquad (E_n - E_s) = (E_n - E_r) + (E_r - E_s)$$

necessarily holds good. This identity may also be written in the form

$$(4) \qquad (E_r - E_s) = (E_n - E_s) - (E_n - E_r).$$

On replacing the energy differences in these equations by the equivalent frequency values, we recognize at once the possibility of deriving new spectral lines from ones already known, by means of *additive or subtractive combinations*. For example, denoting the spectral line emitted on a transition from the five-quantum to the three-quantum orbit by

$\nu_{3,5}$, *etc.*, we can conclude from the given lines $\nu_{3,5}$ and $\nu_{2,3}$ the existence of a line

$$\nu_{3,5} + \nu_{2,3} \; (= \nu_{2,5}) \, .$$

Similarly, given the lines $\nu_{2,3}$ and $\nu_{2,4}$, we can deduce the existence of a line

$$\nu_{2,4} - \nu_{2,3} \; (= \nu_{3,4}) \, ,$$

etc. This so-called *combination principle* was established by *Ritz* in 1908, before Bohr's theory had been developed.

As regards the *intensity of a spectral line*, this obviously depends on how frequently the transition to which it is due occurs.

§ 9. Line Spectrum of Hydrogen.

If, in Bohr's spectral formula, the so-called nuclear charge number (z) be made equal to unity, the formula thus specialized should represent the *line spectrum of hydrogen.* As a matter of fact, complete agreement is found to exist between this purely theoretical formula and the results derived from experimental spectroscopy.

Of the hydrogen lines, those which have been longest

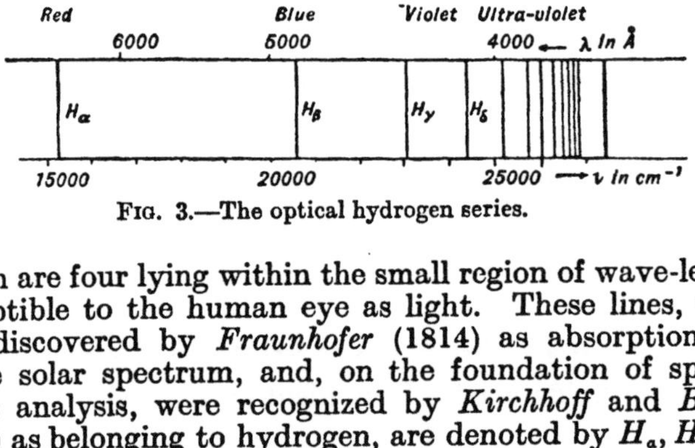

FIG. 3.—The optical hydrogen series.

known are four lying within the small region of wave-lengths perceptible to the human eye as light. These lines, which were discovered by *Fraunhofer* (1814) as absorption lines in the solar spectrum, and, on the foundation of spectroscopic analysis, were recognized by *Kirchhoff* and *Bunsen* (1860) as belonging to hydrogen, are denoted by H_α, H_β, H_γ, and H_δ (in accordance with the chemical symbol for that element). The first and strongest line lies in the red, the second in the blue-green, and the third and fourth in the violet (Fig. 3). They form the first members of a series which continues with decreasing intensity into the ultra-violet, with a limiting wave-length of about 3,600 Å.U.

As long ago as 1885, *Balmer* made the extremely important discovery that very simple numerical relations exist between the wave-lengths of the individual lines in this series. The frequencies of the individual lines can, in fact, be represented by the so-called *Balmer formula*,

$$(1) \qquad \nu = R\left(\frac{1}{4} - \frac{1}{n^2}\right),$$

where n denotes any one of the series of whole numbers from three upwards, whilst R is a constant frequency whose value, as found from the observed frequencies of the lines, is given by

$$(2) \qquad R = 3{\cdot}291 \times 10^{15} \text{ sec.}^{-1}.$$

Since wave-length and frequency are connected by the simple relation

$$(3) \qquad \nu = \frac{c}{\lambda},$$

eqn. (1) may also be written in the form [1]

$$(4) \qquad \frac{1}{\lambda} = \frac{R}{c}\left(\frac{1}{4} - \frac{1}{n^2}\right).$$

Balmer's formula is obeyed with wonderful exactness, and moreover, spectroscopic measurements can be carried out to an extraordinary degree of accuracy, amounting to about a millionth of the measured value. Hence, it has been possible to determine the constant R very exactly from spectroscopic measurements, and the value thus found from the hydrogen spectrum (which fact may be expressed by adding the subscript H) is [2]

$$(5) \qquad \frac{R_H}{c} = 109{,}677{\cdot}69 \text{ cm.}^{-1} (\pm\, 0{\cdot}06).$$

Of the so-called *optical series* of hydrogen, as represented by Balmer's formula, the first twenty lines [3] have been observed in Geissler tubes, and a further thirteen in the spectra of nebulous stars, in which matter is obviously far more attenuated than in even the most highly evacuated Geissler tubes. In the absorption spectrum, however, as many as fifty lines have been detected in laboratory experiments.

It had already occurred to *Ritz*, on the basis of his *com-*

[1] Balmer himself gave his formula the form

$$\lambda = C\,\frac{n^2}{n^2 - 4},$$

where C is a constant equal to $4c/R$.

[2] The value given in eqn. (5) is based on Paschen's measurements.

[3] Thus up to $n = 22$.

bination principle, that, besides the optical series of hydrogen previously described, at least one other series must exist in the *infra-red*, the frequencies of which would be representable by the formula

$$(6) \qquad \nu = R\left(\frac{1}{9} - \frac{1}{n^2}\right), \ (n = 4, 5 \ldots).$$

The first two lines in Ritz's predicted series were actually discovered by *Paschen* in 1909.

It follows from Bohr's theory that the optical series in the emission spectrum arises from a transition into the two-quantum end state, whereas the infra-red series is due to a transition into the three-quantum end state. According to Bohr's theory, we may thus at least expect the existence of a series in the emission spectrum arising from a transition into the one-quantum normal state. This series would be represented by the formula

$$(7) \qquad \nu = R\left(1 - \frac{1}{n^2}\right), \ (n = 2, 3 \ldots).$$

In 1914 this series was actually discovered by *Lyman*, and the first three lines in it are now known. It lies in a somewhat remote region in the *ultra-violet*, and the constant R forms the series limit. Its first line is that due to a transition from the two-quantum state to the one-quantum normal state; and, inasmuch as it arises from the simplest imaginable transition in the simplest of all atoms, this line occupies quite a special position among all the spectral lines of all the elements. Its frequency is equal to three-quarters of R, so that its wave-length is 1215·7 Å.U. Actually, the line gives a very sharp image in all photographic exposures made in the extreme ultra-violet.

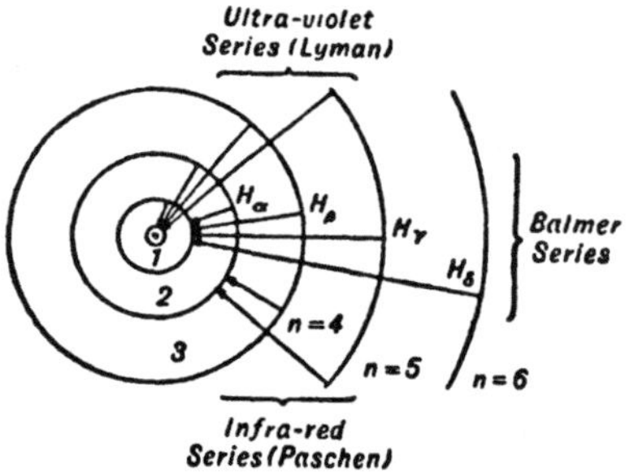

FIG. 4.—Scheme of the hydrogen spectrum.

Finally, in 1922, *Brackett* discovered in the infra-red the first two lines of the series due to a transition to a *four-quantum end state*.[4] (On the same occasion, he also discovered

[4] F. S. Brackett, "*A New Series of Spectrum Lines*," *Nature*, **109** (1922), p. 209. The wave-lengths of the first two lines of the new series are 4·05 μ and 2·63 μ.—An indication of the first line of the series, in the emission spectrum,

the third, fourth, and fifth lines of the Paschen series, of which only the first two lines had up till then been known.)

The origin of the Lyman, Balmer, and Paschen series is diagrammatically represented in Fig. 4, in which the various circular quantum orbits and the transitions from one to another orbit are shown. Another very clear illustration of the origin of the spectral lines is given by the graphical representation of the *energy-levels in the atom* (Fig. 5). By

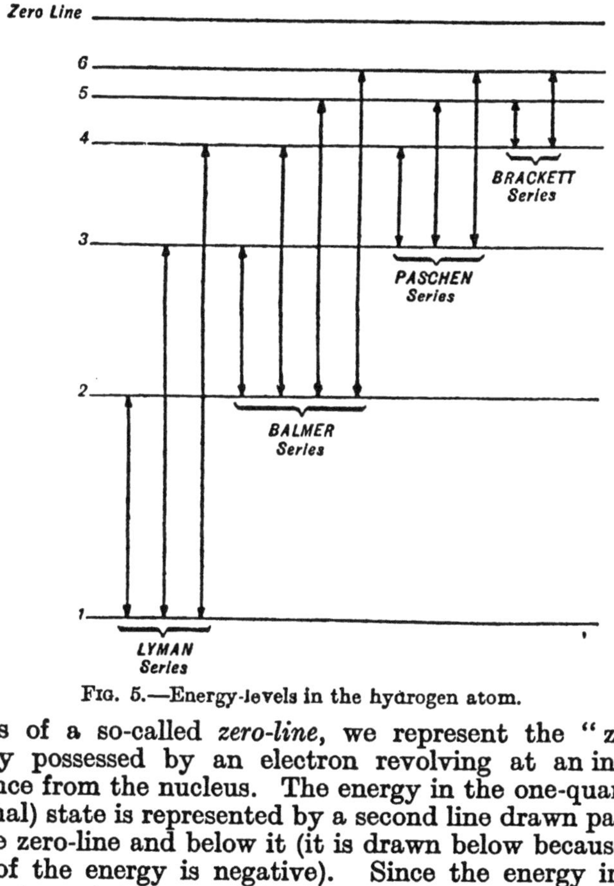

FIG. 5.—Energy-levels in the hydrogen atom.

means of a so-called *zero-line*, we represent the "zero" energy possessed by an electron revolving at an infinite distance from the nucleus. The energy in the one-quantum (normal) state is represented by a second line drawn parallel to the zero-line and below it (it is drawn below because the sign of the energy is negative). Since the energy in the n-quantum state is equal to the normal energy divided by the square of the quantum number, the line representing the two-quantum state lies *between* the one-quantum line and

due to a transition to a five-quantum end state has been found at a wave-length of 74,000 Å. by Pfund [*J. Opt. Soc. Am.* **9** (1924), p. 196].

the zero-line. The line for the three-quantum state lies between that for the two-quantum state and the zero-line, and so on. If we wished our drawing to be to scale, the distances between the individual lines and the zero-line would have to be equal to 1/4th, 1/9th, 1/16th, 1/25th, 1/36th, *etc.*, of the distance between the one-quantum or normal line and the zero-line. In Fig. 5, however, the scale has been intentionally distorted, in order to obtain a clearer diagram.

The *level nearer to the zero-line* always represents the *higher level*, and energy is liberated in every transition which increases the distance from the zero-line, whereas, conversely, energy must be supplied for a transition which diminishes that distance. The method of representation in Fig. 5 is of great advantage, especially in the structurally more complicated atoms. If we denote each level by its corresponding *term* (the energy value divided by h and also, for convenience, by c), the *wave-length* of a spectral line due to a transition between two levels is given very simply by the *reciprocal value of the difference between the terms;* the reciprocal values of the terms themselves represent the *limiting wave-lengths of the various series.* The terms are proportional to the work which must be expended in removing an electron from the orbits to which the individual levels correspond, to beyond the limits of the atom. In the case of the hydrogen atom, the values of the successive terms are simply the quantity R_H/c, as determined by eqn. (5), and 1/4th, 1/9th, 1/16th, 1/25th, 1/36th, *etc.*, of this quantity.

The constant frequency denoted by R is known as *Rydberg's constant*, because, in 1895, Rydberg discovered that this constant has real significance not only for the hydrogen spectrum, but also for the spectra of other elements (especially the alkali metals): thus it represents a *universal physical constant.* Comparing Bohr's theoretically derived formula with the empirical formula established by Balmer, we obtain the relation

$$R = \frac{2\pi^2 e^4 m}{h^3}. \tag{8}$$

This relation, which connects the fundamental constant of spectroscopy with the fundamental quantities of the electron theory and the elementary quantum of action, was first derived by the author of this book in 1910,[5] and more

[5] *Sitz. Ber. der Wiener Akad., Abt. IIa*, 1910, pp. 119–144 [eqn. (67)]; also *Physikal. Zeitschr.*, **11** (1910), p. 537. This derivation, in which the quantum principle was applied for the first time to the theory of the atom, was, however,

accurately later by Bohr from his theory. Eqn. (8), moreover, permits of an exact *determination of the elementary quantum of electricity upon a spectroscopic basis.* For we may write eqn. (8) in the form

$$(9)\qquad R = \frac{2\pi^2 e^2}{\left(\frac{e}{m}\right)\left(\frac{h}{e}\right)^3},$$

and the values of e/m and h/e are known from measurements on the cathode rays and the photo-electric effect respectively. Introducing the values in question, we find for e the same value as follows from the direct determinations of Millikan, although not with the same degree of accuracy.[6]

The product of the Rydberg constant and the elementary quantum of action represents the *energy* (taken with opposite sign) *of the hydrogen atom in the one-quantum* (*normal*) *state,* as follows at once from the representation in terms, and as may also be seen by comparing eqn. (8) with § 7 eqn. (8). Denoting this normal energy by E*, we therefore have

$$(10)\qquad E^* = -hR,$$

and hence

$$(11)\qquad E^* = -2{\cdot}15 \times 10^{-11} \text{ ergs.}$$

We can also readily evaluate the other constants characteristic of the normal state of the hydrogen atom. Giving n and z the value unity in § 7 eqn. (4), we find for the *orbital velocity* of the revolving electron

$$(12)\qquad v^* = \frac{2\pi e^2}{h} = 2{\cdot}188 \times 10^8 \text{ cm. sec.}^{-1}.$$

For the ratio of this velocity to that of light, which represents an important constant, we obtain the value

$$(13)\qquad \alpha = \frac{v^*}{c} = 7{\cdot}29 \times 10^{-3}.$$

We shall denote the *radius of the orbit* in the normal state by a^*. Now the total energy, which (by § 7) represents half the potential energy, is equal to $(-e^2/2a)$, and thus

$$(14)\qquad a^* = -\frac{e^2}{2E^*}.$$

based on Thomson's model of the atom (*vide* § 6), and therefore led to an incorrect numerical factor (instead of the correct factor 2).

[6] This in itself seems to prove that objections to Millikan's determinations are not necessarily at the same time objections to the existence of an elementary quantum of electricity.

Hence it follows from eqn. (11) that

(15) $$a^* = 5{\cdot}30 \times 10^{-9} \text{ cm}.$$

Finally, we obtain the *revolution frequency* in the normal state on dividing the velocity v^* by the circumference of the orbit. We have

(16) $$\omega^* = \frac{v^*}{2\pi a^*},$$

or, by eqns. (12), (14), and (10),

(17) $$\omega^* = 2R;$$

whence we find

(18) $$\omega^* = 6{\cdot}58 \times 10^{15} \text{ sec.}^{-1}.$$

§ 10. Line Spectrum of Ionized Helium.

Whereas the charge on the hydrogen nucleus is a single positive elementary quantum of electricity, in the case of *α-particles* we have become acquainted with positive structures with twice as large a charge and four times as large a mass. Since, on the other hand, the atomic weight of the second lightest element, helium, is four, the idea suggests itself of identifying the α-particle with the *helium nucleus*. A direct confirmation of this idea is found in the fact, which was first established in 1903 by *Ramsay* and *Soddy*, that helium is produced by radioactive substances emitting α-rays.[1]

Therefore, if we put the nuclear charge number z equal to two in Bohr's spectral formula, we might well expect the formula thus specialized to represent the *spectrum of the ionized helium atom*, in which only a single electron revolves round the doubly charged nucleus, the second electron, which is present in the neutral condition, having been removed.

For a transition from an n-quantum to an s-quantum orbit, the frequency produced in ionized helium must be four times as large as that produced in hydrogen, seeing that ν is proportional to z^2. The spectrum of ionized helium would therefore be represented by the formula

(1) $$\nu = 4R\left(\frac{1}{s^2} - \frac{1}{n^2}\right),$$

which may also be written

(2) $$\nu = R\left[\frac{1}{\left(\frac{s}{2}\right)^2} - \frac{1}{\left(\frac{n}{2}\right)^2}\right].$$

[1] A strict proof of this was established somewhat later by Rutherford. For further details, *vide* § 25.

In 1897, *Pickering* actually discovered in the spectrum of the star ζ *Puppis* a series corresponding to eqn. (2) with $s = 4$; moreover, in 1912, *Fowler* discovered in a Geissler tube spectrum a series described by eqn. (2) when s is put equal to 3.[2]

Originally, both the Pickering and the Fowler series were *erroneously ascribed to hydrogen.* Bohr's theory indicates that they are *helium* series, and it was a triumph for that theory when, in 1914, *Paschen* and *Bartels* succeeded in observing the Fowler series in Geissler tubes filled with pure helium and completely uncontaminated by any traces of hydrogen or compounds of hydrogen.

According to the formulæ derived for the line spectra of hydrogen and helium, every second line of the Pickering series should really coincide with a line of the Balmer series,

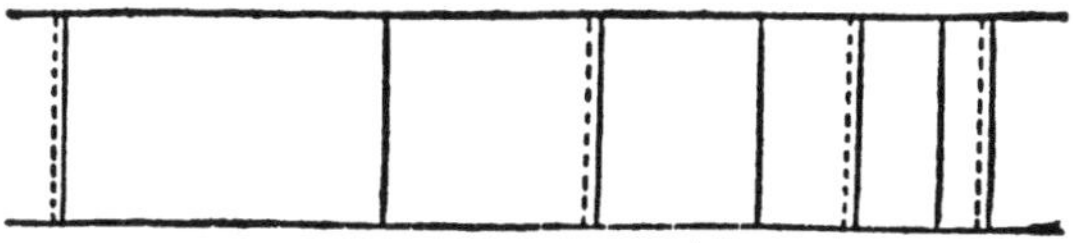

FIG. 6.—Displacement towards the violet of the helium lines.

for the Pickering series is obtained by putting $s = 4$ in eqn. (2) so that, for even values of n, the equation becomes identical with § 9 eqn. (1). Actually, however, the corresponding lines of the Pickering series are found to be slightly displaced in the direction of increasing frequency, and thus towards the violet in the visible region of the spectrum (*cf.* Fig. 6, where the broken lines belong to hydrogen, and the full lines to helium). Although this so-called *displacement towards the violet of the helium lines* might at first appear as a defect in Bohr's theory, it really proved to be a brilliant confirmation of it, when Bohr perfected his original theory by taking the motion of the atomic nucleus into account.

§ 11. The Motion of the Atomic Nucleus.

The previous considerations have been based upon the simple assumption that, in the hydrogen atom and the analogous ionized helium atom, the electron revolves round the nucleus. Some important results, however, followed from a *development of the theory by Bohr* in 1913, when he

[2] Formerly, two Fowler series were distinguished, according as n is odd or even. This distinction, however, appears entirely superfluous from the point of view of Bohr's theory.

took into account the *simultaneous motion of the atomic nucleus.* In reality, the electron does not revolve round the nucleus, but both revolve *about their common centre of gravity.*

Denoting by x and y the respective distances separating the electron and the nucleus from the centre of gravity and by M the mass of the nucleus, we have the ratio

$$x : y = M : m, \quad (x + y = a)\,; \tag{1}$$

whence it follows that

$$x = a\,\frac{M}{M+m}, \quad y = a\,\frac{m}{M+m}. \tag{2}$$

Let the velocity of the electron be v', and that of the nucleus v''. Then, since both electron and nucleus have the same angular velocity, which may be denoted by w,

$$v' = x\,w, \quad v'' = y\,w\,. \tag{3}$$

From this we find the value of the kinetic energy to be given by

$$E_{kin} = \frac{w^2}{2}\,(m\,x^2 + M\,y^2)\,,$$

or, by eqn. (2),

$$E_{kin} = \frac{a^2\,w^2\;M\,m\,(M+m)}{2\,(M+m)^2}\,.$$

In order to simplify the subsequent calculations, let us put

$$\frac{M\,m}{M+m} = \mu\,; \tag{4}$$

then

$$E_{kin} = \frac{a^2\,w^2}{2}\,\mu\,. \tag{5}$$

Again, the centrifugal force must be equal to the force of attraction, so that

$$m\,x\,w^2 = \frac{z\,e^2}{a^2}. \tag{6}$$

Now since, according to eqns. (2) and (4), $m\,x$ is equal to $a\,\mu$, we have

$$\mu\,a^2\,w^2 = \frac{z\,e^2}{a}, \tag{7}$$

and hence, by eqn. (5),

$$E_{kin} = \frac{z\,e^2}{2a}. \tag{8}$$

The potential energy is given by

$$(9)\qquad E_{pot} = -\frac{z e^2}{a},$$

and hence the total energy in the n-quantum orbit is

$$(10)\qquad E_n = -\frac{z e^2}{2a_n}$$

(in agreement with § 7).

For the *angular momentum* we find the value

$$(11)\qquad U = m\,x\,v' + M\,y\,v'' = w\,(m\,x^2 + M\,y^2)\,,$$

so that

$$U = \frac{2E_{kin}}{w},$$

or, by eqn. (5),

$$(12)\qquad U = a^2\,w\,\mu.$$

According to the fundamental quantum relation, the angular momentum multiplied by 2π must be equal to the product of the quantum number and the elementary quantum of action. We therefore have

$$(13)\qquad 2\pi\,a^2\,w\,\mu = n\,h\,.$$

On squaring this equation and dividing by eqn. (7), we obtain

$$(14)\qquad a_n = \frac{n^2\,h^2}{4\pi^2\,z\,e^2\,\mu}\,.$$

Hence, in accordance with eqn. (10), the energy is given by

$$(15)\qquad E_n = -\frac{2\pi^2\,z^2\,e^4\,\mu}{n^2\,h^2}\,.$$

Comparing this formula with § 7 eqn. (8), we recognize a complete agreement, except that m is now replaced by the quantity μ. If we define the frequency R by § 9 eqn. (8), *Bohr's improved spectral formula* becomes

$$\nu = z^2 R \frac{\mu}{m}\left(\frac{1}{s^2} - \frac{1}{n^2}\right),$$

or,

$$(16)\qquad \nu = \frac{R}{1+\frac{m}{M}}\left[\frac{1}{\left(\frac{s}{z}\right)^2} - \frac{1}{\left(\frac{n}{z}\right)^2}\right].$$

Thus, neither in the case of hydrogen nor in that of helium can the constant of the line spectrum be exactly equal to the *Rydberg constant*, as defined by § 9 eqn. (8), in the narrower sense of the term. Rydberg's constant may be denoted by R_∞, because eqn. (16) reduces to the earlier derived spectral formula when the atomic nucleus is assumed to be infinitely heavy in comparison with the revolving electron.

We shall denote the constants of the hydrogen and the helium spectra by R_H and R_{He}, and the nuclear masses of the two atoms by M_H and M_{He} respectively. By eqn. (16) we then have the relations

$$(17)\qquad \begin{cases} R_H = \dfrac{R_\infty}{1+\dfrac{m}{M_H}}, \\[2ex] R_{He} = \dfrac{R_\infty}{1+\dfrac{m}{M_{He}}}, \end{cases}$$

But the mass of the helium nucleus is approximately four times that of the hydrogen nucleus, so that

$$(18)\qquad R_{He} > R_H ;$$

and the explanation of the *displacement towards the violet of the helium lines* is at once evident.

Exact measurements by *Paschen* have given as values for the two quantities R_H and R_{He}

$$(19)\qquad \begin{cases} \dfrac{R_H}{c} = 109{,}677{\cdot}691 \ (\pm\, 0{\cdot}06) \\[2ex] \dfrac{R_{He}}{c} = 109{,}722{\cdot}144 \ (\pm\, 0{\cdot}04), \end{cases}$$

(the value for R_H has already been given in § 9).

Again, the *atomic weights* of hydrogen and helium are

$$(20)\qquad H = 1{\cdot}0077, \quad He = 4{\cdot}00.$$

We may therefore put

$$(21)\qquad \frac{M_{He}}{M_H} = 3{\cdot}97,$$

although, in a more exact calculation, we should also have to take into account the amount contributed toward the total atomic weight by the electron which revolves around the nucleus. Let us put, for brevity,

$$(22)\qquad \frac{m}{M_H} = x$$

Then, by eqn. (17),

$$(23) \qquad \frac{R_H}{R_{He}} = \frac{1 + \frac{x}{3{\cdot}97}}{1 + x}.$$

From this it follows that

$$(24) \qquad x = \frac{R_{He} - R_H}{R_H - \frac{1}{3{\cdot}97} R_{He}},$$

and hence

$$(25) \qquad \frac{1}{x} = \frac{M_H}{m} = \frac{109{,}678 - 27{,}638}{44{\cdot}453} = 1846.$$

From the *displacement towards the violet of the helium lines*, Bohr's theory thus enables us to *calculate* with great accuracy a fundamental constant of the electron theory, *viz.*, the *ratio between the mass of the hydrogen nucleus and the mass of the electron.*

Eqn. (25) may also be written in the form

$$(26) \qquad 1846 = (e/m) : (e/M_H).$$

Now the quotient e/M_H is known from electrolytic measurements (*cf.* § 2), and hence *from the displacement toward the violet of the helium lines* we can also *calculate the specific charge of the electron.* The value thus obtained ($5{\cdot}30 \times 10^{17}$ E.S.U./gram) agrees exactly with that derived from direct measurements on the electric and the magnetic deflection of cathode rays; and this complete agreement naturally constitutes a great triumph for Bohr's theory.

On introducing the value given by eqn. (25) into the first of eqns. (17), we find

$$(27) \qquad \frac{R_\infty}{c} = \frac{R_H}{c} \times 1{\cdot}000542 = 109{,}737{\cdot}1;$$

and hence, by § 9 eqn. (8),

$$(28) \qquad 109{,}737{\cdot}1 \times c = \frac{2\pi^2 e^4 m}{h^3}.$$

Eqns. (26) and (28) attain a still higher degree of accuracy when we also take into account the share contributed towards the atomic mass by the revolving electron. In that case, eqns. (24) and (26) are replaced by more complicated formulæ. In accordance with §§ 1 and 2, we can regard the values of e and of M_H as given with very great accuracy by Millikan's determinations and electrolytic measurements. If we then solve the improved eqns. (26) and (28) with respect to m and h, we obtain for these two quantities also more accurate values than would be possible from measurements on cathode rays, or on the photo-electric effect. As *Flamm* has shown,[1] we find

$$(29) \qquad h = 6{\cdot}545 \times 10^{-27} \text{ erg. sec. } (\pm\, 0{\cdot}012 \times 10^{-27}),$$

$$(30) \qquad \frac{e}{m} = 5{\cdot}304 \times 10^{17} \text{ E.S.U./gram. } (\pm\, 0{\cdot}009 \times 10^{17}),$$

and finally,

$$(31) \qquad m = 9{\cdot}00 \times 10^{-28} \text{ gram. } (\pm\, 0{\cdot}02 \times 10^{-28}).$$

§ 12. Sommerfeld's Theory of Elliptical Orbits.

Two years after Bohr had so successfully evolved his theory of spectra, *i.e.*, in 1915, this theory was the subject of an important extension by *Sommerfeld*, who replaced the circular orbits of the Bohr atomic model by more general *elliptical orbits*, in one *focus* of which the *atomic nucleus* is situated. We know, indeed, from mechanics that, in a periodic motion executed by a body about a centre of attraction under the influence of a force which is inversely proportional to the square of the distance, a *circular orbit* only represents a *special case* of the more general elliptical orbit.

When the considerations, which, based on the quantum theory, enabled Bohr to derive his formulæ for the various quantum states of the hydrogen atom, are now correspondingly modified and generalized, they lead to the following result (the process of reasoning is too complicated to be repeated here) :—

For the *semi-major axis* of the orbital ellipse, the same manifold and the same values are obtained as were found in Bohr's original theory for the radii of the circular orbits. Denoting the semi-major axis by a and the radius of Bohr's circular orbit by r, we have the relation

$$(1) \qquad (a_n)_{\text{SOMMERFELD}} = (r_n)_{\text{BOHR}},$$

where for r_n is to be substituted the right-hand side of

[1] *Physikal. Zeitschr.*, **18** (1917), p. 515.

§ 7 eqn. (3). With every value of the major axis as thus determined, however, there may now be associated a *series* of possible values for the semi-minor axis (b), according to the simple formula

$$b_{n,k} = a_n \frac{k}{n}, \quad (k = 1, 2 \ldots n). \tag{2}$$

In Sommerfeld's theory, therefore, each orbit requires two *quantum numbers* for its definition: a "*principal quantum number*" (n), and a "*subordinate quantum number*" (k), which, however, can never be larger than the principal number. We shall speak symbolically of a state n_k, or of an n_k-orbit, thereby writing the subordinate number as subscript to the principal number.

Contrasted with Bohr's theory, Sommerfeld's theory thus points to a *greater variety* of possible states. In Bohr's theory, there is only a simple series of states distinguished by the numbers

1, 2, 3, 4, 5, *etc.*

In Sommerfeld's theory there exists a greater manifold, characterized by the numerical symbols

1_1; $2_1, 2_2$; $3_1, 3_2, 3_3$; $4_1, 4_2, 4_3, 4_4$; $5_1, 5_2, 5_3, 5_4, 5_5$; *etc.*

The *principal quantum number* determines the *length* of the major axis; the *subordinate number* the *eccentricity*, and therefore the *shape* of the orbit. The smaller the subordinate number, for a given principal number, the more eccentric

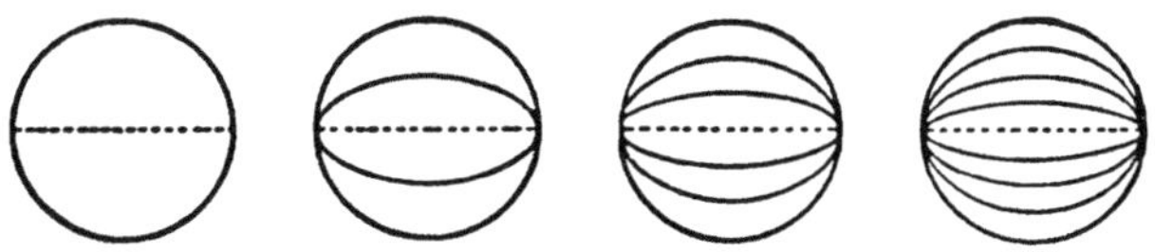

FIG. 7.—Orbital ellipses in accord with the quantum theory.

and elongated is the ellipse. If the *subordinate number is equal to the principal number*, the orbit is *circular*, and we have the special case on which the original Bohr theory was based.

In Fig. 7 the possible forms of the ellipses for a given principal quantum number are represented; for the sake of greater distinctness the ellipses are drawn concentrically, whereas the actual arrangement is confocal. In Fig. 8 this confocal arrangement is reproduced, but on different scales for different values of the principal number: for the sake

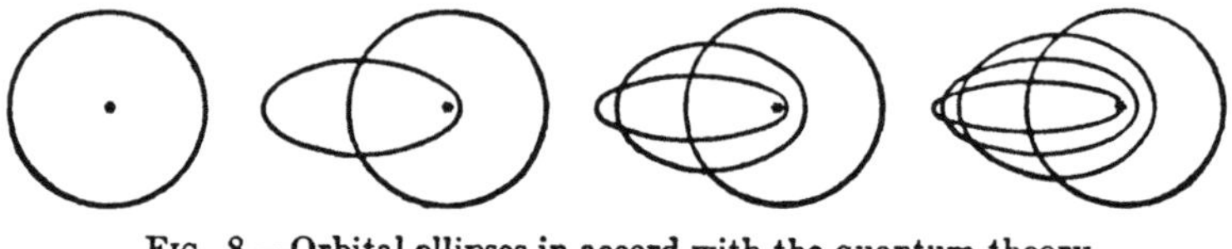

FIG. 8.—Orbital ellipses in accord with the quantum theory.

of clearness, the major axis has been given the same length throughout, although it is really proportional to the square of the principal quantum number. This fact is taken into account in Fig. 9, where we see reproduced on the proper scale the elliptical orbits for $n = 1$ to 4.

If n and s are two principal numbers, it follows from Sommerfeld's theory that a total of $n \times s$ *transitions* between the n- and s-*quantum* states are conceivable. This number, however, is considerably *reduced* by the so-called *selection principle*, which was introduced by *Rubinowicz* in 1918, and was also derived by *Bohr* as a special case of a more general principle (the "correspondence principle"). The selection principle states that *only those transitions are permissible* in which the *subordinate number varies by unity.*[1]

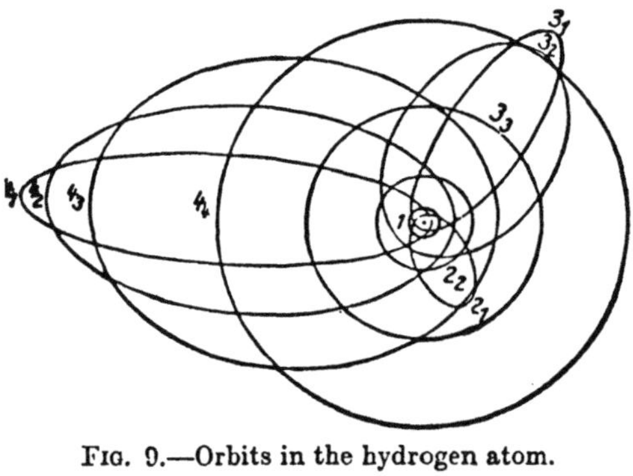

FIG. 9.—Orbits in the hydrogen atom.

Thus, for example, if we consider the states whose principal number is s, we can at once see that a total of $(2s - 1)$ regular transitions from states with an arbitrary principal number n will lead to these states, where $n > s$. For s_1 can be combined only with n_2, s_2 only with n_1 or n_3, s_3 only with n_2 or n_4, *etc.* Hence, only a single transition leads from any

[1] For further details of the selection principle and apparent exceptions thereto, as well as of the correspondence principle, *vide* Sommerfeld's "*Atombau und Spektrallinien,*" ch. v. (English translation by H. L. Brose: London, Methuen), and Bohr, "*Three Essays on Spectra and Atomic Constitution,*" 2nd essay (Cambridge University Press).

higher quantum state to the one-quantum state; three transitions lead from any higher to the two-quantum state; five from a higher to the three-quantum state, and so on; whereby the one-quantum state consists of a single modification, the two-quantum of two, and the three-quantum state of three modifications, *etc.* Each line in the Lyman series can therefore be due only to one definite transition, whereas each line in the optical Balmer series can be produced by three, and each line in the infra-red Paschen series by five transitions. According to the selection principle, a transition between two circular orbits is only possible when the principal numbers differ by unity; thus, only the first lines in all series can arise from a transition between two circular orbits.

As regards the *energy* possessed by an electron in an elliptical orbit, calculation (too complicated to be repeated here) leads, in the first place, to the result that it *depends only on the principal quantum number.* The value given by this calculation for the energy in an n-quantum orbit is identical with that resulting from Bohr's theory of circular orbits [§ 7 eqn. (8)], irrespective of the value of the subordinate number. The correctness of the results obtained in Bohr's theory is therefore in no way affected by Sommerfeld's development.

Thus, although the above calculation leads at first to the result that, for a given value of n, the energy in all the variously shaped orbits is the same, yet the calculation is based on the essential assumption that the elliptical motion of the electron under consideration is not disturbed in any way. On the other hand, if any kind of *perturbation* be present, the influence of the subordinate number must be felt, whether these perturbations be due to other electrons likewise revolving about the nucleus (which is always the case, except with hydrogen and ionized helium), or whether they be due to the establishment of an external electric or magnetic field. In these cases (which will be more fully discussed later), Sommerfeld's hypothesis of a subordinate quantum number proves to be extraordinarily fruitful. For in all these cases, the energy values appertaining to the same principal number must *differ* from each other; hence, in accordance with Bohr's *frequency condition,* a so-called *resolution of the spectral lines* must take place. No longer does a transition from an n-quantum to an s-quantum state give rise to a single line, as in the original theory which did not concern itself with subordinate numbers, but a *group* of more or less closely associated lines is produced.

It was, however, an especial triumph for the hypothesis of the subordinate quantum number that, in consequence of the different eccentricities of the orbits, the *spectral lines* even in the undisturbed *hydrogen atom* must be resolved. This resolution, which was predicted by Sommerfeld and which has also been completely confirmed by quantitative experiments, is simply a result of the minute *correction* to be applied to the above-mentioned calculations, when, in contradistinction to classical mechanics, the *theory of relativity* is taken into account. Whereas classical mechanics leads to the result quoted above, according to which the energy for a given principal number is independent of the subordinate number, a certain degree of dependence is nevertheless found to exist when, in accordance with the theory of relativity, the increase in mass with increasing velocity is taken into account; for, of course, the velocities differ in the variously shaped elliptical orbits.

§ 13. The Fine Structure of Spectral Lines.

When, in accordance with § 3 eqn. (6), the *dependence of the mass upon the velocity* demanded by the theory of relativity is taken into account, a complicated calculation (which cannot be repeated here) leads to the result that, in the first place, a steady *rotation of the perihelion of the ellipse*, slow compared with the period, is superimposed upon the motion in the ellipse (Fig. 10). Furthermore, we find for the *energy* in the state of the hydrogen atom (or hydrogen-analogue) characterized by the principal number n and subordinate number k the expression

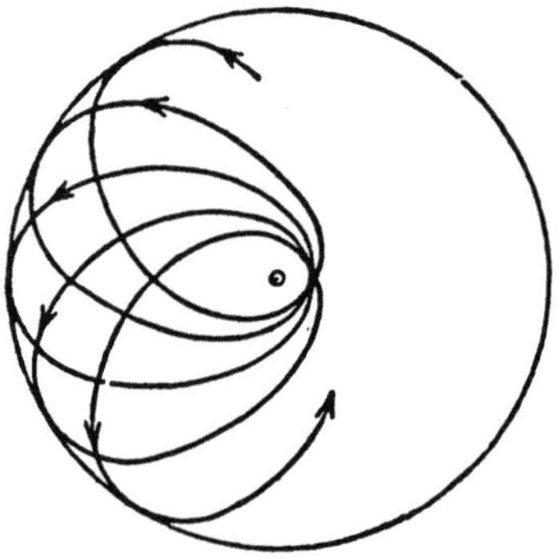

Fig. 10.—Perihelion rotation in the orbital ellipses.

$$(1) \qquad E_{n,k} = -\frac{R\,h\,z^2}{n^2} + \frac{3R\,h\,z^4\,\alpha^2}{4n^4} - \frac{R\,h\,z^2\,\alpha^2}{n^3\,k}.$$

In this expression R, as usual, denotes Rydberg's constant, h the elementary quantum of action, z the nuclear charge number, and α the ratio of the velocity in the normal hydrogen orbit to the velocity of light.

We see from eqn. (1) that the value of the energy is composed of three terms. The first term has the same value as E_n in Bohr's theory. The second and third terms are *relativistic correction terms*, but only the third depends also on the subordinate number k. Thus, by eqn. (1), to every principal quantum number n there correspond n *different energy-levels*. The larger k is, the smaller is the final correction term in eqn. (1), and since the sign of this term is negative as is the sign of the energy, a *larger subordinate number* corresponds to a *higher energy-level* (because of the smaller absolute magnitude of the energy). Fig. 11 gives a schematic representation of the energy-levels, although, to make it clearer, the scale has been distorted.

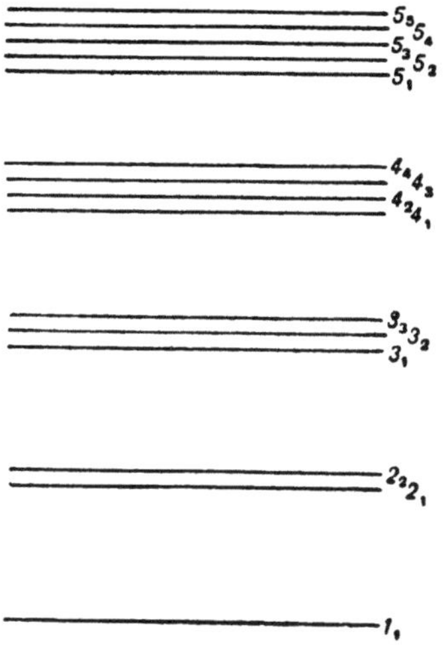

FIG. 11.—Scheme of energy-levels.

The manifold of energy values is, of course, associated with a manifold of *spectral terms*. The frequency difference of the two spectral terms thus corresponding to the two-quantum state of the hydrogen atom is known as the *frequency difference of the hydrogen doublet* ($\Delta\nu_H$). This quantity is therefore equal to the difference between the values of the last term in eqn. (1) divided by h (with $z = 1$ and $n = 2$), when k is put equal to one or to two respectively. Hence

$$(2) \qquad \Delta\nu_H = \frac{R\alpha^2}{2^3}\left(\frac{1}{1} - \frac{1}{2}\right),$$

or,

$$(3) \qquad \Delta\nu_H = \frac{R\alpha^2}{16}.$$

Substituting for α the value given in § 9 eqn. (13), and for R the value given in § 11 eqn. (27), we find

$$(4) \qquad \frac{\Delta\nu_H}{c} = 0{\cdot}365 \text{ cm.}^{-1}.$$

The term differences in the other quantum states stand, of course, in simple numerical relations to the frequency difference of the hydrogen doublet. If, for example, we consider the *three-quantum* state, we find from eqn. (1) for the spectral terms (denoted by T) corresponding to its modifications

$$(5)\qquad T_{3,3} - T_{3,2} = \frac{R\alpha^2}{3^3}\left(\frac{1}{2} - \frac{1}{3}\right) = \frac{8}{81}\Delta\nu_H ,$$

and

$$(6)\qquad T_{3,2} - T_{3,1} = \frac{R\alpha^2}{3^3}\left(\frac{1}{1} - \frac{1}{2}\right) = \frac{8}{27}\Delta\nu_H ,$$

so that

$$(7)\qquad T_{3,3} - T_{3,1} = \frac{32}{81}\Delta\nu_H .$$

Let us now consider the hydrogen line H_α, *i.e.*, the first line of the optical series, which arises by a transition from the three-quantum to the two-quantum state. In accordance with the selection principle, there are three regular possible ways in which this transition may occur, *viz.*,

$$\begin{aligned}(\mathrm{I})\quad & 3_2 \longrightarrow 2_1 ,\\ (\mathrm{II})\quad & 3_3 \longrightarrow 2_2 ,\\ (\mathrm{III})\quad & 3_1 \longrightarrow 2_2 .\end{aligned}$$

For the frequency differences of the spectral lines we have then the relations (*cf.* Fig. 12)

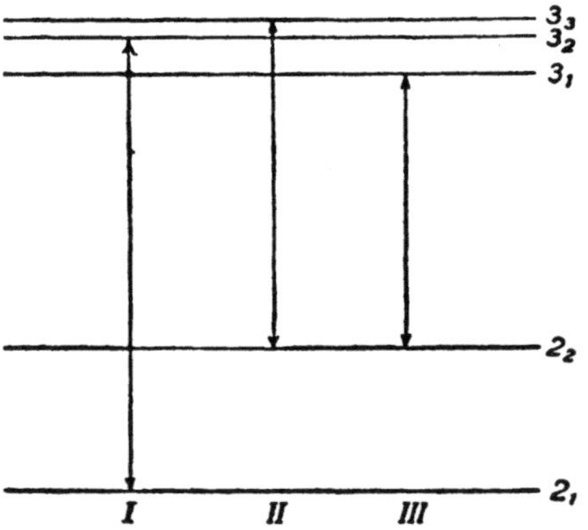

FIG. 12.—Fine structure of the H_α-line.

$$(8)\begin{cases} \mathrm{I} - \mathrm{II} = \frac{73}{81}\Delta\nu_H ,\\ \mathrm{II} - \mathrm{III} = \frac{32}{81}\Delta\nu_H \\ \mathrm{I} - \mathrm{III} = \frac{35}{27}\Delta\nu_H .\end{cases}$$

Considerations exactly similar to those used as an illustration in the case of the H_α-line may, of course, also be applied to all the other hydrogen lines. Moreover, all formulæ which hold for a hydrogen line that arises by a transition from an n-quantum to an s-quantum state obviously hold for a helium line due to a similar transition, provided that the quantity $\Delta\nu_H$ be replaced by the quantity $\Delta\nu_{He}$, which is defined in a precisely analogous manner. The last term in eqn. (1) shows us that, on account of the proportionality to the fourth power of the nuclear charge number,

(9) $$\Delta\nu_{He} = 16\,\Delta\nu_{H}\,,$$

so that, by eqn. (3),

(10) $$\Delta\nu_{He} = R\,\alpha^2\,.$$

Now, in the case of hydrogen, the observation and measurement of the fine structure is extremely difficult, because the intervals between the lines are so small. In the visible region of the spectrum, the difference in wave-length corresponding to the quantity $\Delta\nu_H$ is only about one-twentieth to one-fifth of 1 Å.U.[1] In the case of helium, however, conditions are much more favourable to observation, because, in accordance with eqn. (10), the separation for helium is 16 times as large as for hydrogen (the values of the initial and final quantum numbers being the same), so that the interval between neighbouring lines is considerably larger.

Whereas in the Balmer series the lines only appear double even in spectroscopic apparatus of high resolution (simply because stronger resolution is not available), it has been possible to measure accurately the individual lines as resolved in the spectrum of ionized helium. In this way the value of $\Delta\nu_{He}$ could be experimentally determined, and the value of $\Delta\nu_H$ followed, as it is one-sixteenth thereof. Measurements by *Paschen* gave as a result

(11) $$\frac{\Delta\nu_H}{c} = 0{\cdot}3645\ \text{cm.}^{-1}\ (\pm\ 0{\cdot}0045)\,.$$

The complete agreement between the theoretically calculated and the experimentally determined values constitutes a brilliant triumph not only for Bohr's and for Sommerfeld's conceptions, but also for the theory of relativity, the mass formula of which is splendidly confirmed by this agreement.[2]

[1] It follows from the elementary relation

$$c = \lambda\nu$$

that $$\Delta\lambda = -\frac{c}{\nu^2}\Delta\nu = -\lambda^2\frac{\Delta\nu}{c}.$$

On giving λ a value of between 4×10^{-5} and 8×10^{-5} cm., and $\Delta\nu$ the value in eqn. (4), we find for $\Delta\lambda$ a value of from 5 to 20×10^{-10} cm.

[2] Successful attempts have recently been made to measure the fine structure in the hydrogen spectrum itself; the results confirm Sommerfeld's theory completely. [*Cf.* G. M. Shrum, *Proc. Roy. Soc.*, *A*. **105** (1924), p. 259.] In this connection, attention may be drawn to the fact that it is possible to determine the constants *e*, *h*, and *m* solely from spectroscopic measurements, if Faraday's electrochemical constant be assumed as given. For that purpose, three equations in the theory of spectra are available, *viz.*, § 11 eqns. (26) and (28), and § 13 eqn. (11) [in combination with § 13 eqn. (3).]

§ 14. The Stark Effect.

The wonderful regularities of the hydrogen spectrum show most clearly that all atomic phenomena are governed by *simple integral relations*, but this fact finds even plainer expression in the *changes* produced in the *hydrogen spectrum* by an *external electric field*.

The resolution of spectral lines in an electric field was discovered by *Stark* in 1913, and is therefore generally known as the *Stark effect* (Fig. 13).[1] The *theory* of the effect was developed from the *quantum principle* by *Schwarzschild* and *Epstein* in 1916. Between the results of theory and experi-

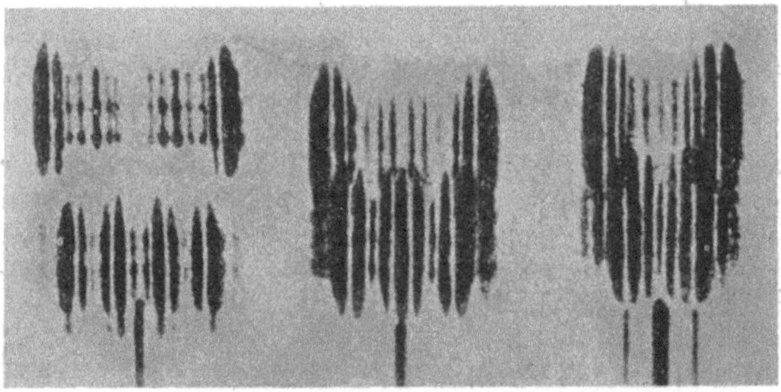

FIG. 13.—Stark effect in the hydrogen lines H_δ, H_γ, and H_β.

ment such an excellent agreement was found to exist that the Stark effect must be regarded as one of the most brilliant empirical confirmations of the quantum principle. The quantum theory of the Stark effect is, however, too difficult to be described in detail here, and we shall only state and discuss the result derived from the theory. According to this, the frequency (ν) of a hydrogen line in an electric field of strength (E) is given by the expression

$$(1) \qquad \nu = \nu_0 \pm z\, E\, \sigma,$$

where ν_0 is the frequency of the line in the absence of an external field, and σ is a universal constant of the form

[1] The discovery of the Stark effect only succeeded after the skilful surmounting of great experimental difficulties. The observation of the resolution of the lines is, in fact, only possible in strong fields, which cannot be produced in a Geissler tube, because it is too good a conductor of electricity. Stark therefore used canal ray particles to produce the spectral lines to be resolved, while he introduced a subsidiary electrode into the immediate neighbourhood of the cathode emitting the canal rays, thereby producing a very large potential difference over the short interval of a few millimetres.

$$(2) \qquad \sigma = \frac{3h}{8\pi^2 e m},$$

h denoting the elementary quantum of action, e and m the charge and mass of an electron. The "*ordinal number*" z, which is always an *integer*, depends on the principal quantum numbers n and s which characterize the initial and final states on the emission of the spectral line ; it is given by the relation

$$(3) \qquad z = nn' \pm ss', \begin{cases} n' = 0, 1, 2 \ldots (n-1), \\ s' = 0, 1, 2 \ldots (s-1). \end{cases}$$

The numbers n' and s' can thus assume all integral values between zero (and including it) and n and s (but excluding them) respectively.

The theory also leads to a rule governing the direction of vibration in the light corresponding to the spectral lines. If the expression

$$(4) \qquad (n - s) + (n' - s') = l$$

is an *even* number, the *electric vibrations* are *parallel* to the direction of the electric field-strength. On the other hand, if l is an *odd* number, the electric vibrations will occur *transversely* to the direction of the field-strength. There are thus *two different classes of components* to be distinguished, which may be called the p- and the t-components from the initial letters of the words "parallel" and "transverse."

Now, as is well known, light waves are always propagated only in directions normal to the direction of vibration. Hence, if we are observing the resolution in the direction of the lines of force—the so-called *longitudinal effect*—the p-components will not be visible, and only the t-components will be seen in an unpolarized state. If, however, we observe the resolution at right angles to the lines of force—the so-called *transverse effect*—all components will be visible, but the p- and the t-components will be readily distinguishable owing to their different polarization.

The results of the theory may be discussed by reference to two simple *examples, viz.*, the *two violet hydrogen lines* H_γ and H_δ. As has already been mentioned, the line H_γ is produced as third line of the Balmer series by a transition from a five-quantum to a two-quantum state. Hence, for the line H_γ, $n = 5$ and $s = 2$, so that n' can assume the values 0, 1, 2, 3, 4, but s' can only be 0 or 1. The ordinal number z is therefore equal either to $5n'$ or to $(5n' \pm 2)$. In the first case the number l, as defined by eqn. (4), is equal to $(3 + n')$; in the second case, it is equal to $(2 + n')$. The ordinal numbers of the p-components, for which l must be even, would thus be $5n'$ for odd values of n', and $(5n' \pm 2)$ for even values. Hence the p-components of the line H_γ would have the ordinal numbers

$$2, 5, 8, 12, 15, 18, 22.$$

On the other hand, the ordinal numbers of the t-components, for which l must be odd, would be $5n'$ for even values of n', and $(5n' \pm 2)$ for odd values. Hence the t-components of the line H_γ would have the ordinal numbers

$$0, 3, 7, 10, 13, 17, 20.$$

The complete *agreement* between theory and experiment is to be seen in Fig. 14, in which are reproduced the results of Stark's observations, the length of each line representing the intensity of the component in question, as estimated by Stark from the image on the photographic plate.

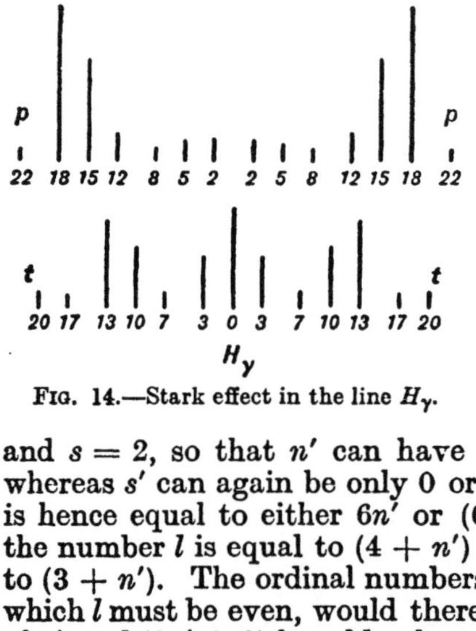

FIG. 14.—Stark effect in the line H_γ.

The fourth line in the Balmer-series, H_δ, arises on a transition from a six-quantum to a two-quantum state. For this line we have thus to put $n = 6$ and $s = 2$, so that n' can have the values 0, 1, 2, 3, 4, 5, whereas s' can again be only 0 or 1. The ordinal number z is hence equal to either $6n'$ or $(6n' \pm 2)$. In the first case the number l is equal to $(4 + n')$; in the second it is equal to $(3 + n')$. The ordinal numbers of the p-components, for which l must be even, would therefore be $6n'$ for even values of n', and $(6n' \pm 2)$ for odd values. Hence the p-components of the line H_δ would have the ordinal numbers

$$0, 4, 8, 12, 16, 20, 24, 28, 32.$$

On the other hand, the ordinal numbers of the t-components, for which l must be odd, would be $6n'$ for odd values of n', and $(6n' \pm 2)$ for even values. The t-components of the line H_δ would therefore have the ordinal numbers

$$2, 6, 10, 14, 18, 22, 26, 30.$$

As may be seen from Fig. 15, the agreement between theory

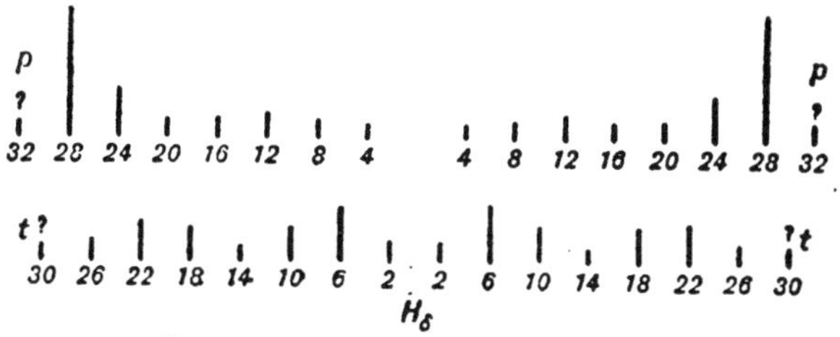

Fig. 15.—Stark effect in the line H_δ.

and the previously known experimental results is excellent also in the case of the line H_δ.[2]

§ 15. The Zeeman Effect.

Just as in an electric field, a *resolution of spectral lines* also occurs in a *magnetic field*. Considerations of the quantum theory, into which we cannot enter here, show that the value of the energy, in an arbitrary orbit, possessed by an electron in a magnetic field is given by

(1) $$E = E_0 + k' \frac{e h}{4\pi m c} H.$$

In this expression E_0 denotes the energy in the absence of a magnetic field, H the magnetic field-strength, and k' a whole number which, as follows from the quantum theory, can only change by unity or not at all as a result of a transition. In the latter case, Bohr's frequency condition shows that, for an arbitrary spectral line,

(2) $$\nu = \nu_0,$$

where ν_0 is the frequency in the undisturbed atom. On the other hand, if k' changes by ± 1, it follows from the frequency condition that

(3) $$\pm (\nu - \nu_0) = \frac{e}{4\pi m c} H.$$

Just as in the Stark effect, the resolution in a magnetic field is also proportional to the field-strength. There is, however, an essential difference between the two effects, for the magnetic resolution gives rise only to two lines in addition to the original line, and it is completely independent of all quantum numbers, and therefore of the initial and the final

[2] The occurrence of the components with the highest ordinal numbers is, it is true, doubtful in the experimental results, whilst the line of zero ordinal number is absent, possibly because its intensity is too small.

state. Even the elementary quantum of action itself is lacking in eqn. (3), for it disappeared from the calculation on the application of the frequency condition.

It also follows on the basis of the quantum theory that the emitted spectral line must be *circularly polarized* when k' changes by unity, but *plane polarized* when k' remains unchanged after a transition. In that event, the linear electric vibrations take place in the direction of the magnetic lines of force, so that, if the resolution be observed in this direction, the component ν_0 will be missing. Hence, in the *longitudinal effect* a *doublet* must be visible, but in the *transverse effect* a *triplet*. It follows from the theory that the two outer components must be *circularly polarized in opposite senses*, and in such a way that the sense of polarization in the component of smaller wave-length must be the same as the sense of the electric current which would produce, within a solenoid, a magnetic field of the same direction as the field applied.

The resolution of spectral lines in a magnetic field was experimentally discovered by *Zeeman* in 1896, on which account this phenomenon is commonly known as the *Zeeman effect*. In the case of spectral lines which, in the absence of a magnetic field, are simple, complete quantitative agreement is found to exist between the results of experiment and theory. In the case of originally *multiple* lines, however, considerably more complicated resolution phenomena occur, and these are known as the *anomalous Zeeman effect*, which cannot be discussed here.[1]

Immediately after Zeeman's discovery, *Lorentz* developed a *magneto-optical theory* which, since it was evolved four years before the quantum theory, was of course in no way concerned with the elementary quantum of action and quantum numbers. Lorentz's theory was simply based on the assumption that the electrons, to which spectral lines are due, are bound to a definite position of equilibrium by quasi-elastic forces.[2] Although this conception has been super-

[1] For a further discussion of the anomalous Zeeman effect *vide*, for example, Sommerfeld's "*Atomic Structure and Spectral Lines*," Ch. vi., § 7; the theory of the normal Zeeman effect is to be found in Ch. v., § 6.—The resolutions observed in the Zeeman effect are considerably smaller than those of the Stark effect. Thus, the resolution in the strongest magnetic fields only amounts to about one Å.U., whereas, in an electric field of 30,000 volt/cm., for example the resolution of the blue hydrogen line (H_β) extends over 13 Å.U.—A very interesting application of the Zeeman effect is to astrophysical problems, and by its aid *Hale* discovered strong magnetic fields in the *sun*.

[2] Forces are termed quasi-elastic when they are proportional to the distance from a position of equilibrium. *Cf.* § 6, note 2.

seded by the later developments of atomic physics, Lorentz's theory nevertheless led to eqn. (3) and to the correct rule of polarization. Eqn. (3), as already obtained by Lorentz, thus rendered possible the first determination of the ratio e/m, for measured values of the resolution and of the magnetic field-strength. Moreover, the negative character of the electronic charge was rendered evident in view of the polarization rule, by the observed sense of the circularly polarized lines. Since the quotient e/m proved to be more than a thousand times as large as in the case of the ionized hydrogen atom, the Zeeman effect first revealed to physicists the existence of particles considerably lighter than the lightest of the atoms, up till then regarded as indivisible.

CHAPTER III

RÖNTGEN RAYS

§ 16. Origin of Röntgen Rays.

WHEN the motion of *rapidly moving electrons* is *checked*, the resulting phenomenon represents a *reversal of the photoelectric effect*. The kinetic energy of corpuscular radiation is then partially transformed into *electromagnetic wave* energy; and the arrest of a rapid motion of electrons thus produces the remarkable rays discovered by *Röntgen* in 1895, and known as *Röntgen rays* or X-rays.

The most striking property of Röntgen rays is their extraordinary *power of penetration*. They are scarcely absorbed at all by objects which are completely opaque to ordinary light. Among their other properties, particular mention should be made of their strong ionizing power, and the fact that they experience no deflection whatever either in a magnetic or in an electric field; it thus appeared that Röntgen rays could not be of a corpuscular nature. The penetrating power, or *hardness*, was soon found to be a reliable characteristic for *distinguishing* between various kinds of Röntgen rays; the greater or smaller the power of penetration, the harder or softer is the relevant radiation said to be.

In Röntgen's original experiments, the newly discovered rays originated in those portions of the glass wall of a cathode ray tube which were bombarded by the cathode rays.[1] The cathode ray tube gradually developed into a *Röntgen ray tube* when, on the one hand, a so-called *anti-cathode, viz.*, a plate of some heat-resistant metal (platinum or tungsten), was introduced into the interior of the tube to serve as emission centre of the Röntgen rays, and, on the other hand, a suitable shape was given to the cathode so as to concentrate the cathode rays at a point on the anti-cathode.[2] It

[1] The discovery was due to Röntgen's observation that a preparation of barium platinocyanide (which fluoresces strongly under the influence of cathode rays) also becomes luminous in the neighbourhood of a cathode tube at places which are out of reach of the rectilinearly propagated cathode rays.

[2] For this purpose the cathode is shaped like a concave mirror. It is necessary to concentrate the cathode rays at a point in order that the Röntgen rays thus generated may produce sharp images.

was soon discovered that the hardness of the Röntgen rays increases with increasing tube potential, and, in tubes of modern construction,[3] it can be regulated at will by means of the potential.

The absence of any electric or magnetic deflection in itself led physicists to suspect that Röntgen rays represent electromagnetic waves. This concept found strong support in the demonstration, afforded by *Barkla* in 1905, that the *secondary Röntgen radiation* emitted by a body upon which primary Röntgen rays impinge is partially *polarized*.[4] The *transverse nature* of Röntgen rays therefore appeared proved.

Moreover, the relationship between Röntgen rays and light must also appear obvious from purely theoretical considerations. Röntgen rays are produced when quickly moving electrons lose their velocity. Now, since an electron is surrounded by an electromagnetic field so long as it is in motion, but only by an electrostatic field from the moment that it comes to rest, it follows at once that the transition to a state of rest, or so-called retardation, must give rise to the *propagation of an electromagnetic disturbance*, a so-called electromagnetic *pulse*. Having recognized this fact, *Stokes* and *Wiechert* in 1896 evolved a theory of the origin of Röntgen rays, in which they regarded the radiation emitted by the anti-cathode as essentially an *impulse radiation.*

If we now make the plausible assumption that the electron is slowed down along the portion of the path traversed *inside an atom*, we find for the "width of the pulse" (corresponding to the wave-length) a value of the order of 10^{-9} cm., or about one ten-thousandth of the wave-length of visible light. Furthermore, Einstein's law, which governs the mutual transformation of corpuscular and undulatory radiation, leads to the same order of magnitude for the wave-length of Röntgen radiation. If, for example, we take the potential (V) of a Röntgen tube as being 100,000 volts (= approx. 300 E.S.U.), and equate the products $e\,V$ and $h\,\nu$, we obtain a value of the order of 10^{19} sec.$^{-1}$ for the frequency. This value is again about ten thousand times as large as the frequency of visible light.

[3] In the Röntgen ray tubes devised by *Lilienfeld* and *Coolidge* this is effected essentially by making an incandescent wire the source of the electrons, and the vacuum is so high that a true discharge is no longer possible. The chief difficulty connected with the use of the older tubes, *viz.*, that the hardness of the Röntgen radiation depended on the gas pressure, is thus avoided.

[4] This was proved by the fact that the intensity of the tertiary rays, emitted by a body bombarded by the secondary rays, varied with the position relative to the body.

The assumption of so small a wave-length finds additional support in the failure of all early experimental efforts to demonstrate *interference* or *diffraction* phenomena with Röntgen rays, by means of any ordinary optical method.[5] The usual diffraction gratings could not be applied, for, as follows from the theory of the grating (and as we shall show in detail in the next section), the interval between two lines of the grating must not be much larger than the wave-length of the light investigated. Hence, researches on Röntgen rays would require gratings upon which there were millions of lines to a millimetre. The technical impossibility of preparing such fine gratings seemed to exclude almost every prospect of physicists ever being able to determine the exact wave-lengths of Röntgen rays, until, in 1912, *Laue* conceived the ingenious idea of making use simply of *crystals*, instead of artificial gratings, for diffracting the rays. Laue's idea was based on a theory developed by the mineralogist *Bravais* about 1850, according to which the regular form of crystals must be due to a *regular arrangement of the atoms in the crystal.* It also followed from the known density of crystals, and from the known value of the Loschmidt number, that the mutual distances between the atoms in crystals must be of the order of 10^{-8} cm. Provided that the individual atoms behave as diffraction centres, it was therefore to be expected that *crystals* would actually *behave as gratings when used with Röntgen rays.*

§ 17. Space-Lattices.

In order to understand the behaviour of crystals with respect to Röntgen rays, we shall first consider, as the simplest of all gratings or lattices, a *line-lattice* in which the lattice points are arranged at equal intervals along a straight line. The constant interval between two successive lattice points is called the *lattice constant.*

We now suppose a *wave* to be incident on the lattice, the direction of the ray making with the direction of the lattice an angle whose cosine is α_0.[1] The individual lattice points may be regarded as diffraction centres, each of them consti-

[5] The first diffraction experiments with Röntgen rays were made by Haga and Wind in 1899. Their experiments were repeated by Walter and Pohl in 1908, using an improved method, and Koch's careful photometric measurements, later carried out on their exposures, enabled Sommerfeld in 1912 to calculate from the very slight indication of diffraction a width of about 4×10^{-9} cm. for the predominant pulse. At the present time, however, these experimental results are only of historical interest.

[1] α_0 is therefore the sine of the angle of incidence.

tuting a centre of excitation of new waves. These waves travel in all directions, but it will be immediately evident that they will be *diffracted with especial intensity in those directions*, for which the *difference in path* of the rays proceeding from two successive lattice points amounts to an *integral multiple of a wave-length*. In fact, more accurate theoretical considerations (which we cannot discuss here) show that, in consequence of mutual *interference* between the diffracted rays, an *appreciable* diffraction of the light occurs only in the directions just given.

We shall now denote by α the cosine of the angle included between the direction of the lattice and one of the rays thus diffracted. Then, if a be the lattice constant, the difference

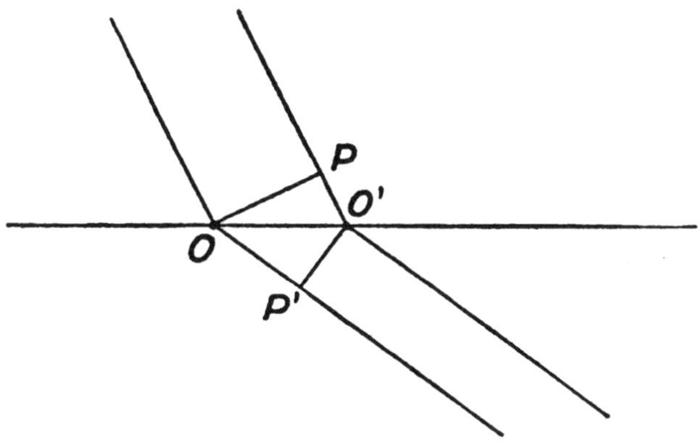

FIG. 16.

in path of the rays diffracted in this direction by two neighbouring lattice points is equal to $a\,(\alpha - \alpha_0)$, as is obvious from Fig. 16. (If O and O' be the two lattice points, and $O\,P$ and $O'\,P'$ the perpendiculars dropped on the incident and diffracted rays respectively, the difference in path is equal to the difference between the intervals $O\,P'$ and $O'\,P$.) Rays of given wave-length λ are therefore diffracted only in those directions (with a cosine α), for which the relation

(1) $$a\,(\alpha - \alpha_0) = n\,\lambda$$

is satisfied for *integral* values of the so-called *ordinal number* (*order*) n. Conversely, if the incident radiation is composed of rays of all wave-lengths, it is *resolved into spectra* by the diffraction, a different diffraction spectrum corresponding to each value of the ordinal number n. (In passing it may be mentioned that, if we turn the diffracted rays about the direction of the lattice in Fig. 16, we obtain the directions of the light reflected by diffraction, though, of course, it is not a case of reflection in the true sense of the word.)

It follows from eqn. (1) that diffraction can never occur, unless

$$(2) \qquad a > \lambda .$$

Moreover, a must not be *much* larger than λ, because otherwise, in the spectra of low order which are of particular importance, the difference $(\alpha - \alpha_0)$ would be too small and the spectral resolution insufficient.[2]

From the one-dimensional line-lattice, let us now pass to the two-dimensional *plane-lattice*, which will be formed by points with the coordinates

$$(3) \qquad x = k_1 a, \quad y = k_2 b$$

constructed in a plane coordinate system, where a and b are two lattice constants, and k_1 and k_2 can assume every integral value, either positive or negative. The coordinate axes may be inclined at any angle to each other, but we shall limit ourselves to a rectangular system in which the two lattice constants are identical (Fig. 17).

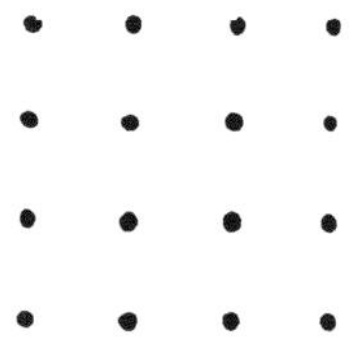

FIG. 17.

Let us suppose rays to impinge on such a plane-lattice, making with the x- and the y-axis angles whose cosines are α_0 and β_0. Eqn. (1) then expands into a system of two equations, *viz.*,

$$(4) \qquad \begin{cases} a(\alpha - \alpha_0) = n_1 \lambda , \\ a(\beta - \beta_0) = n_2 \lambda , \end{cases}$$

where α and β are the cosines of the angles included between the diffracted ray and the x- and the y-axis. Since the two equations are mutually independent, they can be solved with respect to the unknowns α and β. Hence, to every value of the wave-length there corresponds, for a given direction of incidence and given ordinal numbers n_1 and n_2, a definite direction of diffraction determined by the values of α and β [according to eqn. (4)]. Thus a *plane-lattice* also leads to a resolution into spectra of the incident radiation, this resolution forming a *dual manifold of diffraction spectra.*[3]

[2] In the best optical line-gratings (*Rowland* concave gratings), a is always less than ten times the wave-length.

[3] Diffraction phenomena of this kind are to be observed when a street lamp is viewed through a thin umbrella.

We accomplish the transition from the plane-lattice to *space-lattices* by imagining a three-dimensional coordinate system in a space filled with points whose coordinates are

$$(5) \qquad x = k_1 a, \quad y = k_2 b, \quad z = k_3 c\,;$$

where a, b, c are lattice constants, and the numbers k_1, k_2, k_3 assume every integral value, positive and negative. When the coordinate system is rectangular, we term the space-lattice *orthogonal.* In that case, if the three lattice constants a, b, c are all different, we define the space-lattice as being *orthorhombic;* we call it *tetragonal,* when two of the three lattice constants are identical, and *regular* or *cubic* when all three have the same value. If only two of the three coordinate axes are at right angles to each other, the space-lattice is said to be *monoclinic;* and if all three axes are inclined to each other, it is termed *triclinic.*

The following considerations are limited to the special case of a *cubic* space-lattice, in which, therefore, the axes are orthogonal and the lattice constants identical. Let rays meet such a space-lattice in a direction including with the three coordinate axes angles whose cosines are $\alpha_0, \beta_0, \gamma_0$. In place of eqns. (4), we then have the following *triplet of equations* :—

$$(6) \qquad \begin{cases} a(\alpha - \alpha_0) = n_1 \lambda\,, \\ a(\beta - \beta_0) = n_2 \lambda\,, \\ a(\gamma - \gamma_0) = n_3 \lambda\,. \end{cases}$$

There is, however, an essential difference between this triplet of equations and eqn. (4). In eqn. (4) we were able to regard as unknowns two magnitudes which completely determined the direction of the diffracted ray for a given direction of incidence, for given wave-length and order; the number of unknowns was the same as the number of equations. In the present case, however, there are only two unknowns in the three equations (6), if we take again as given the direction of incidence, wave-length, and order. For the three directional cosines α, β, γ determining the direction of the diffracted ray are not mutually independent; two of these cosines are always connected with the third in accordance with the elementary relation

$$(7) \qquad \alpha^2 + \beta^2 + \gamma^2 = 1\,.$$

Since we cannot regard the ordinal numbers as unknowns, inasmuch as they can only assume perfectly definite, permissible values, *viz.,* integral values (and, indeed, only the lower values require consideration), it appears that, in the three equations, the *wave-length* itself must also be an *unknown* for a given direction of incidence. It is not, indeed, limited to a single value, but it may have any of a *series of possible values* corresponding to the different possible integral values of n_1, n_2, n_3.

These values of the wave-lengths can easily be found from eqns. (6) and (7). We write eqns. (6) in the form

$$(8) \qquad \begin{cases} \alpha = \alpha_0 + n_1 \dfrac{\lambda}{a}, \\ \beta = \beta_0 + n_2 \dfrac{\lambda}{a}, \\ \gamma = \gamma_0 + n_3 \dfrac{\lambda}{a}. \end{cases}$$

We then square these equations and add them together, taking due account of eqn. (7) and the analogous equation connecting the directional cosines $\alpha_0, \beta_0, \gamma_0$. In this way we find

$$1 = 1 + 2\frac{\lambda}{a}(n_1\alpha_0 + n_2\beta_0 + n_3\gamma_0) + \frac{\lambda^2}{a^2}(n_1^2 + n_2^2 + n_3^2)$$

or,

$$(9) \qquad \lambda = -2a\frac{n_1\alpha_0 + n_2\beta_0 + n_3\gamma_0}{n_1^2 + n_2^2 + n_3^2}.$$

In contrast to line- and plane-lattices, a *space-lattice* thus gives *no continuous diffraction spectrum;* its diffractive effect, on the contrary, is to *single out from the incident radiation* those rays whose *wave-lengths* satisfy eqn. (9) for *integral* values of n_1, n_2, n_3.

Eqns. (6) also enable us readily to calculate the *angle* included between the *diffracted* and the *incident ray*. Denoting this angle by 2θ, we have [by an elementary theorem of solid geometry]

$$(10) \qquad \cos 2\theta = \alpha\alpha_0 + \beta\beta_0 + \gamma\gamma_0.$$

On squaring eqns. (6) and adding, we find [bearing eqn. (7) in mind]

$$a^2(2 - 2\cos 2\theta) = \lambda^2(n_1^2 + n_2^2 + n_3^2).$$

But, in accordance with a well-known trigonometrical formula,

$$(11) \qquad \sin\theta = \sqrt{\frac{1 - \cos 2\theta}{2}}.$$

Hence,

$$\sin\theta = \frac{\lambda}{2a}\sqrt{n_1^2 + n_2^2 + n_3^2}. \tag{12}$$

From the diffracting lattice point O, we cut off an interval OA one unit in length in the direction of the incident ray produced, and an equal interval OB in the direction of the diffracted ray (Fig. 18). We then construct the *median plane* MM' between the incident and the diffracted ray, by bisecting the angle AOB and erecting through the bisector a plane at right angles to the plane AOB. Every point on the median plane must therefore be equidistant from A and B.

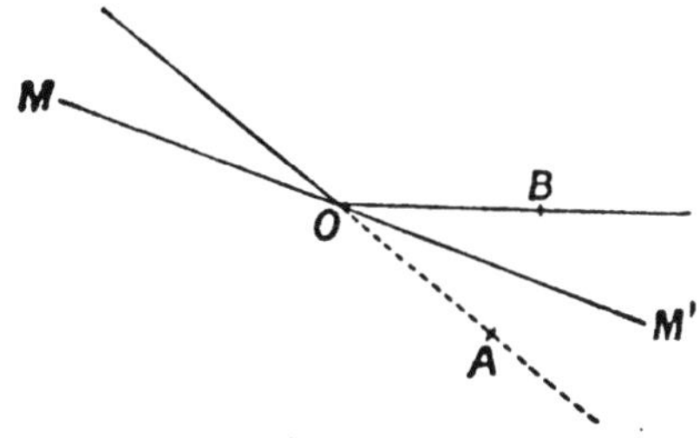

FIG. 18.

With the point O as origin, we next construct a coordinate system with axes parallel to the original axes. With respect to this system, the coordinates of the point A are $\alpha_0, \beta_0, \gamma_0$, while those of the point B are α, β, γ. Since an arbitrary point (x, y, z) on the median plane is equidistant from A and B, it must satisfy the relation

$$(x-\alpha_0)^2 + (y-\beta_0)^2 + (z-\gamma_0)^2 = (x-\alpha)^2 + (y-\beta)^2 + (z-\gamma)^2. \tag{13}$$

Bearing in mind eqn. (7) and the analogous one for $\alpha_0, \beta_0, \gamma_0$, we find therefore

$$x(\alpha-\alpha_0) + y(\beta-\beta_0) + z(\gamma-\gamma_0) = 0, \tag{14}$$

or, by eqns. (8),

$$x n_1 + y n_2 + z n_3 = 0. \tag{15}$$

Referred to a lattice point as origin, this is the *equation of the median plane* between the incident ray and the ray diffracted at that point, the order of which is determined by the whole numbers n_1, n_2, n_3.

§ 18. Lattice Planes.

In a cubic lattice, let us suppose intervals to be cut off along the three coordinate axes from the origin, these intervals being equal to the quotients of the lattice constant and three whole numbers k_1, k_2, k_3, which have no common denominator. The extremities of the three intervals then define a plane, whose equation we wish to determine.

Since the coordinates of any arbitrary plane are interconnected by a *linear* relation, the most general form of the equation of any *plane* is

(1) $$D_1 x + D_2 y + D_3 z = D_4 ,$$

where the four magnitudes denoted by D are constants. Let

(2) $$D_4 = D_1 C_1 = D_2 C_2 = D_3 C_3 .$$

Eqn. (1) then becomes

(3) $$\frac{x}{C_1} + \frac{y}{C_2} + \frac{z}{C_3} = 1 .$$

If now we consider, for instance, the point where the plane cuts the x-axis, both y and z are to be equated to zero for this point of intersection, and we see therefore that the three constants C_1, C_2, C_3 simply represent the intercepts of the plane upon the three coordinate axes. The plane which we considered above cuts off an interval of magnitude a/k, along the x-axis, where a is the lattice constant. Hence the equation of this plane is given by

(4) $$k_1 x + k_2 y + k_3 z = a .$$

In general, we *define as a lattice plane of the space-lattice* every plane for which the left-hand side of eqn. (4) is equal to an *integral multiple of the lattice constant;* as the general equation of a lattice plane, we have therefore

(5) $$k_1 x + k_2 y + k_2 z + s a ,$$

where s denotes a whole number. Since, however, s can assume the values of each one of the whole series of integers, the three so-called *indices of the lattice planes* (k_1, k_2, k_3) determine a *set of parallel planes*, one of which, *viz.*, the one for which $s = 0$, passes through the origin.[1] For the rest, s may of course be either positive or negative.

It is also easy to see that an *infinite number of lattice points* will be distributed in a reticular manner over every lattice plane, which is supposed to be unlimited in extent. If m be any integer, positive or negative, and we put

(6) $$\begin{cases} x = y = m k_3 a , \\ \quad z = - m (k_1 + k_2) a , \end{cases}$$

then

$$k_1 x + k_2 y + k_3 z = 0 .$$

[1] For the point ($x = y = z = 0$) satisfies eqn. (5) when $s = 0$.

That is to say, all lattice points represented by eqns. (6) lie on the lattice plane ($s = 0$) which passes through the origin. The same thing obviously holds good for all lattice points which are represented by the equations to be obtained from eqns. (6) by a cyclic interchange, *viz.*,

$$y = z = m\, k_1\, a\,, \quad x = -\, m\, (k_2 + k_3)\, a\,,$$

or,

$$z = x = m\, k_2\, a\,, \quad y = -\, m\, (k_3 + k_1)\, a\,.$$

Let us now consider any arbitrary plane of the set of lattice planes, for which the value of s differs from zero, and let us form, for instance, the *indeterminate equation*

$$(7) \qquad k_1 x' + k_2 y' = s\,.$$

Naturally this equation always has integral solutions when both x' and y' may be negative: let x^* and y^* be such a pair of solutions. If then we consider the lattice point

$$(8) \qquad x = x^* a\,, \; y = y^* a\,, \; z = 0\,,$$

we find for this point, according to eqn. (7),

$$k_1 x + k_2 y + k_3 z = s\, a\,;$$

i.e., the lattice point determined by eqn. (8) belongs to the lattice plane in question. We have therefore only to remove the origin to this lattice point in order to be able to apply the proof advanced above to the lattice plane under consideration, and thus to show that it also contains an infinite number of lattice points.

The *distance between two successive lattice planes* in a set determined by a triplet of indices may be denoted by d. This is obviously equal to the *perpendicular* dropped from the origin upon the lattice plane for which $s = 1$. This plane cuts off intercepts a/k_1, a/k_2, a/k_3 along the three coordinate axes. Denoting by φ_1, φ_2, φ_3 the angles included between the perpendicular and the three axes, we have

$$(9) \qquad d = \frac{a}{k_1} \cos \varphi_1 = \frac{a}{k_2} \cos \varphi_2 = \frac{a}{k_3} \cos \varphi_3\,.$$

But the sum of the squares of the three cosines is equal to unity. Hence it follows from eqns. (9) that

$$(10) \qquad d = \frac{a}{\sqrt{k_1^2 + k_2^2 + k_3^2}}\,.$$

On comparing eqn. (5) with § 17 eqn. (15), we recognize, first, that the *median plane* between the incident ray and the ray diffracted in a space-lattice always represents a *lattice plane*. We see further, however, that the indices of this lattice plane always correspond to the quotients of the ordinal numbers, determining the order of the diffracted ray, and their highest common factor. We have

$$n_1 = n k_1, \quad n_2 = n k_2, \quad n_3 = n k_3, \tag{11}$$

where n is the highest common factor of the ordinal numbers n_1, n_2, n_3. The diffraction of the incident rays may therefore be regarded as a *reflection at the lattice planes* of the space-lattice, the simple relation expressed by eqn. (11) connecting the ordinal numbers of the diffraction with the indices of the reflecting lattice planes.

The angle made by the incident ray with the lattice plane, *i.e.*, the complement of the angle of incidence, is called the *glancing angle*. In accordance with eqn. (11) and § 17 eqn. (12), it is determined by the relation

$$\sin\theta = \frac{n\lambda}{2a}\sqrt{k_1^2 + k_2^2 + k_3^2}. \tag{12}$$

In virtue of eqn. (10), however, this may also be written

$$2d\sin\theta = n\lambda, \tag{13}$$

where n is, of course, an arbitrary integer.

If now we consider two lattice planes, the distance (AB) between which is equal to d (Fig. 19), the difference in path between two rays reflected at the points A and B will be equal to $(CB + BD)$, where C and D are the feet of the perpendiculars dropped from the point A on the directions of the ray reflected by the second plane. But $(CB + BD)$ is equal to $2d\sin\theta$. Hence, a set of lattice planes *singles out* from the incident radiation those rays for which an *integral multiple of the wave-length* is *equal to the difference in path* between two rays, which are reflected from two successive lattice planes.

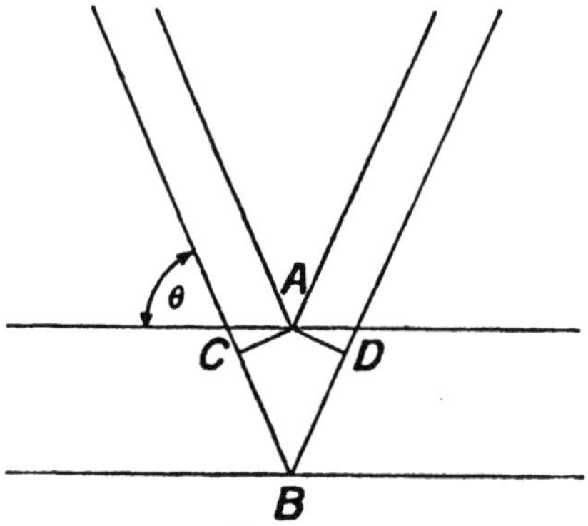

FIG. 19.

As a matter of fact, this law [which is, of course, only another way of expressing § 17 eqn. (9)] is very plausible. For, if the diffraction of the rays takes place in such a way that they appear to be reflected at the lattice planes in accordance with the usual laws of optics, then the reflected rays must strengthen and weaken each other owing to interference. Moreover, an appreciable increase in intensity will only be observed when the difference in path between the rays reflected at two successive planes amounts exactly to an integral multiple of a wave-length.

Furthermore, as we have shown in the previous considerations from two entirely different points of view, a space-lattice only diffracts incident rays of quite definite wave-length from radiation with a given direction of incidence. Thus complete diffraction of a narrow bundle of parallel rays can obviously only be attained if the *space-lattice be rotated*. A complete *resolution into spectra* is then actually attainable, as may be seen from some simple geometrical relations.

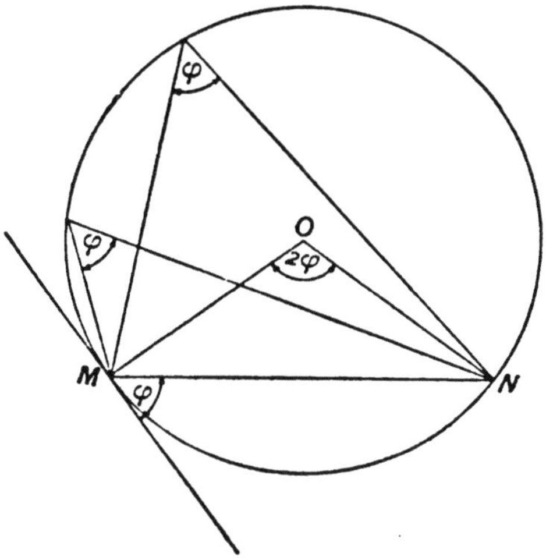

FIG. 20.

In order to derive these relations, we start from a well-known elementary theorem, which states that all the *peripheral angles subtended by the same chord in a circle* are *equal in magnitude*, and also equal to the *tangential angle* included between the chord and the tangent to the circle, drawn at one end of the chord (Fig. 20).[2]

Let us now consider (Fig. 21) a *lattice plane*, and suppose a narrow bundle of rays, passing through a slit R, to meet it at the point O with a glancing angle θ. Denoting the position of the plane by the straight line $A\,A'$, we therefore have

$$\angle\, R\,O\,A' = \theta . \tag{14}$$

[2] The second part of the theorem, which is also well known, is most simply proved from the fact that the angle at the centre is twice the angle at the circumference. Hence in Fig. 20 the angle $M\,O\,N$ is equal to 2ϕ, so that the angle $O\,M\,N$ is equal to half of $(180 - 2\phi)$ The tangential angle, however, is complementary to the angle $O\,M\,N$, because the radius is perpendicular to the tangent. Thus the tangential angle is equal to ϕ.

We draw a circle with $R\,O$ as a chord, and the line $A\,A'$ touching it tangentially at O.

Let us now *rotate* the space-lattice about an axis perpendicular to the plane of the figure, so that the lattice plane turns from the position $A\,A'$ into the position $B\,B'$. Let the straight line $B\,B'$ intersect the circle at the point P. Then the angle $R\,P\,O$, a peripheral angle subtended by the chord $R\,O$, is, according to the theorem just quoted, equal to the tangential angle ROA'. Thus we also have

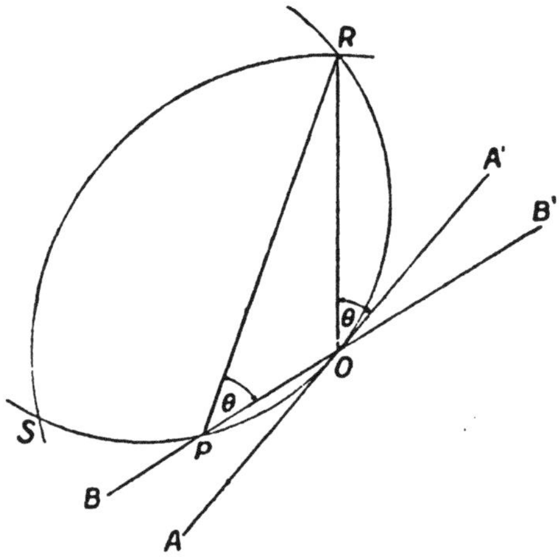

FIG. 21.

$$(15) \qquad \angle\, R\,P\,O = \theta\,.$$

By eqn. (13), however, a definite value of the glancing angle is associated with a definite value of the wave-length, for a given ordinal number n.

When the space-lattice is *rotated*, the point in the lattice plane at which the reflection of a ray of definite wave-length takes place thus *travels along the circumference of the circle* which, passing through the points R and O, touches at O the straight line $A\,A'$ that makes an angle θ with the line $R\,O$. While the lattice plane is being rotated, the ray of definite wave-length thus *glides* along it.

We now suppose (Fig. 21) a *second circle* to be described about O as centre, with $O\,R$ as radius. Let this second circle cut the first circle $R\,O\,P$ at a point S. It may be easily shown that every ray which strikes the lattice plane at a glancing angle θ must, after reflection, pass through the point S, whatever may be the position of the lattice plane at the instant of reflection.

We see (Fig. 22) that the angle $O\,S\,R$, a peripheral angle subtended by the chord $O\,R$, is equal to the angle $R\,P\,O$; hence, by eqn. (15),

$$(16) \qquad \angle\, O\,S\,R = \theta\,.$$

But $S\,O\,R$ is an isosceles triangle, and it follows therefore from eqn. (16) that the angle at the apex of this triangle

$$(17) \qquad \angle SOR = 180° - 2\theta.$$

Thus, by eqn. (14),

$$(18) \qquad \angle SOA = \theta;$$

i.e., the ray reflected at the point O, when the lattice plane is in the position AA', passes through the point S after reflection.

Moreover, the angles SOR and SPR are peripheral angles subtended by the same chord SR. Hence by eqn. (17) we have

$$(19) \qquad \angle SPR = 180° - 2\theta,$$

so that, by eqn. (15),

$$(20) \qquad \angle SPB = \theta.$$

Therefore, the ray reflected at the point P, when the lattice plane is in the position BB', likewise passes through the point S after reflection. Furthermore, since the position BB' is an entirely *arbitrary* position of the lattice plane, every ray proceeding from the point R, which is reflected at an angle θ from the lattice plane passing through O, passes through the point S after reflection. For a given order of diffraction, however, a quite definite value of the wave-length corresponds to each glancing angle. Hence, on rotation of the space-lattice, all rays of the same wave-length are, as it were, "*focussed*" at one and the same point.

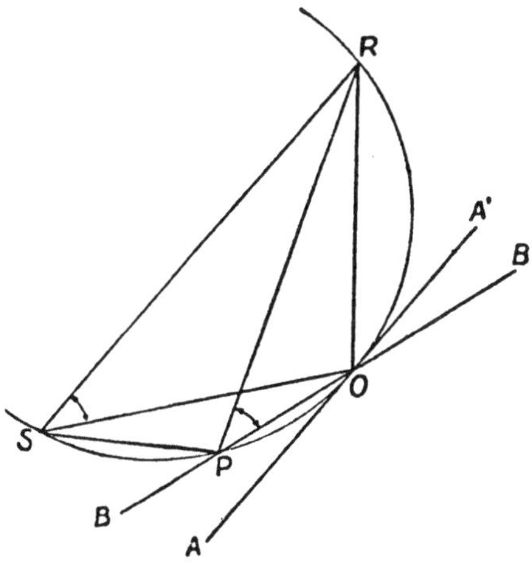

Fig. 22.

§ 19. The Interference of Röntgen Rays in Crystals.

If *crystals* possess the structure of *space-lattices*, and their lattice constants, though larger, are not much larger than the wave-lengths of *Röntgen rays*, these rays should be *diffracted by interference* within the crystals. This diffraction will follow the laws derived from purely geometrical considerations in the previous sections, for the *atoms* corresponding to the *lattice points* must obviously act as *diffraction centres* with respect to electromagnetic waves.

An *experimental proof* of an *interference of Röntgen rays* produced by *crystals* would therefore, on the one hand, demonstrate the lattice structure of crystals, and, on the other hand, confirm the conjecture that Röntgen rays are electromagnetic waves with a wave-length of the order of 10^{-9} cm.

Starting out with this idea, *Laue*, in 1912, was the first to investigate the action of crystals on Röntgen rays, and actually to demonstrate experimentally the suspected phenomena. In these researches, which were carried out by Laue in collaboration with *Friedrich* and *Knipping*, a narrow parallel pencil of rays was separated from the radiation of a Röntgen tube by means of lead screens with adjustable slits. The rays fell on a plate of crystal about 1 cm. square and $\frac{1}{2}$ mm. thick, and a photographic plate was mounted some centimetres behind the crystal.

Now, from a parallel bundle of incident rays, a stationary space-lattice diffracts rays only in quite definite directions, which, apart from the direction of incidence, depend only on the ordinal numbers of the diffraction. [These directions are defined by § 17 eqns. (8), in which λ is to be given the value following from § 17 eqn. (9).] The points at which the directions of these diffracted rays *meet* the plane of the photographic plate must thus be *blackened* on the plate.

Experiment actually gave these results. In Figs. 23 and 24 (on Plate I) are reproduced the so-called *Laue photographs* which were obtained with a crystal of *zinc blende* (ZnS). In the exposure shown in Fig. 23, the distance of the plate from the crystal amounted to 3·5 cm., whereas it was only 1 cm. in the exposure reproduced in Fig. 24. On that account, the spots appear more closely packed in the second photograph.

In both exposures, the rays met a *face of the cube at right angles*. Their direction was therefore that of an edge of the cube, and this explains the *fourfold symmetry* to be observed in the photographs. An axis of symmetry in a crystal is termed n-fold, when through this axis n planes can be drawn, with respect to which the crystal is symmetrical. Now it is easy to see that each *edge of a cube* in a regular crystal repre-

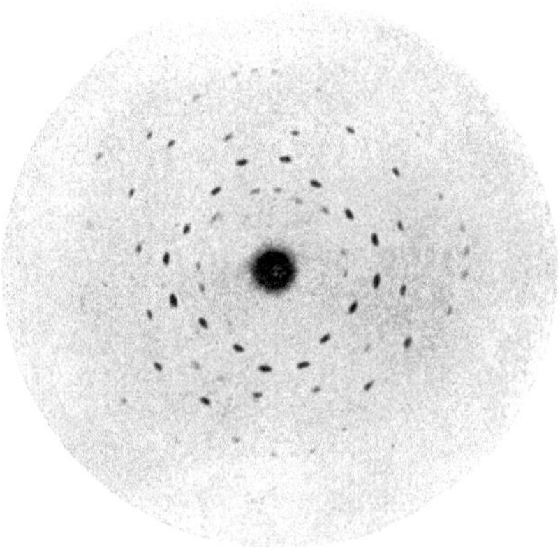

FIG. 23.

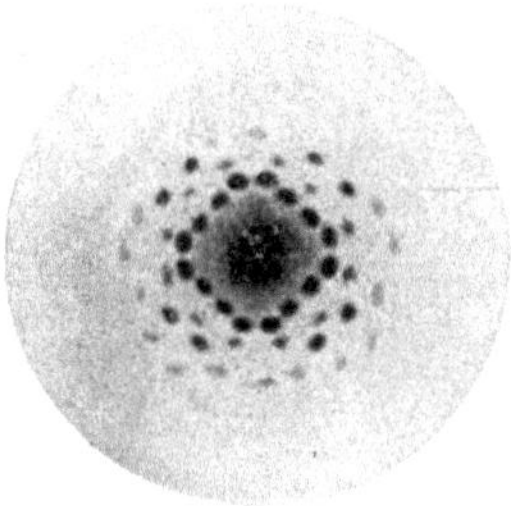

FIG. 24.

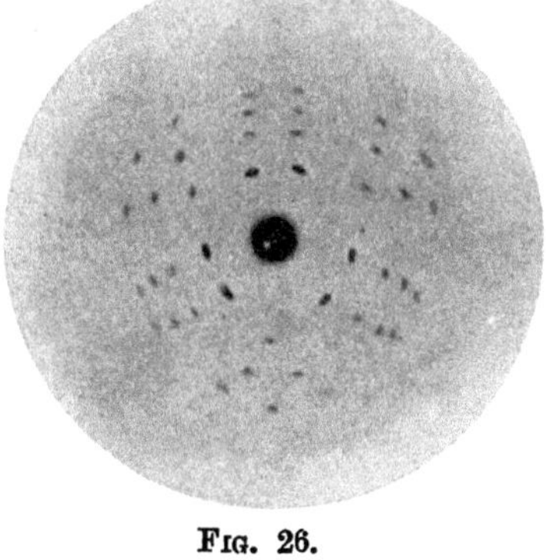

FIG. 26.
Laue Photographs.

[To face p. 68.

sents a *fourfold axis of symmetry*. For, through each edge, four planes of symmetry can be drawn, the lines along which they intersect a plane normal to the edge being represented in Fig. 25; two of these intersecting lines are again edges, and two face diagonals. Thus the photograph exhibit a fourfold symmetry, provided that the incident pencil of rays be perpendicular to a face of the cube. Corresponding to Fig. 25, four lines of symmetry are to be recognized, and hence all spots (provided they do not lie just on a line of symmetry) are repeated *eight times* in all. This is actually to be seen in Figs. 23 and 24.

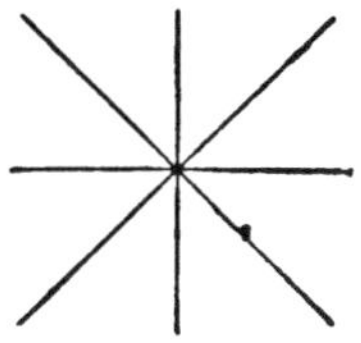

FIG. 25.

If the ray falls perpendicularly on an *octohedral* face cut from a regular crystal, geometrical considerations (into which we cannot enter here) show that its direction is that of a trigonal axis of the cube—a threefold axis of symmetry. An exposure made in this way is reproduced in Fig. 26 (on Plate I), in which each spot occurs *six times*.

In 1913, *W. H.* and *W. L. Bragg* (father and son) deduced from the Laue photographs most important conclusions on the *structure of crystals*. Although the space-lattice nature of crystals had been proved beyond doubt, yet, before crystals could be actually utilized for accurate measurements of wave-lengths, the exact lattice structure of at least *one* crystal had to be determined, and in this connection there were two questions to be solved.

The first question concerns a possible *centring of the space-lattice*. A cubic lattice may be simple, space-centred, or finally, face-centred. The symmetry relations are the same in all three cases. A cubic lattice (such as was considered in §§ 17 and 18) is *space-centred* when, in all the elementary cubes whose edges are given by the lattice constant, an additional lattice point is situated at the centre of the cube; each elementary cube then contains nine instead of eight lattice points. If, on the other hand, the interior of the cube is unoccupied, whilst lattice points are situated at the centres of the six faces (Fig. 27), a *face-centred* lattice results.

FIG. 27.

The second fundamental question of crystal structure arises from the fact that most crystals are composed of *different kinds of atoms*, *e.g.*, rock-salt is made up of sodium and chlorine atoms. It might well be expected that each kind of atom would form a separate space-lattice, but the question was still unsolved as to how the different space-lattices were combined in the crystal.

Very ingenious reasoning, which we cannot discuss here,[1] enabled W. H. and W. L. Bragg to answer these questions for numerous crystals, by investigating the Laue photographs. The arrangement and number of the interference spots differ, according as to whether and how the space-lattice is centred. Moreover, the relative positions of the combined space-lattices of the various kinds of atoms could be determined, on the grounds of the fact that the behaviour of an atom with respect to Röntgen rays depends on the number of electrons contained in the atom, and thus, broadly speaking, on the *atomic weight.* Hence a lattice plane of atoms reflects Röntgen rays in a different manner according as the two species of atom have much the same, or widely differing atomic weights. Thus, in spite of a presumably identical crystal structure, the different alkali halides, for example, should give rather different Laue photographs. [*E.g.*, in potassium chloride (sylvine) the two atomic weights only differ slightly (39 and 35·5), but in potassium bromide they are very different (39 and 80).] A comparison of the photographs of various alkali halides thus rendered possible a recognition of the structure of crystals of these compounds. In a later section we shall discuss at length the general results of crystal analysis; for the moment, we shall only refer to the structure of the alkali halides, because one of them, rock-salt (NaCl), is the principal crystal used in the measurement of the wave-lengths of Röntgen rays.

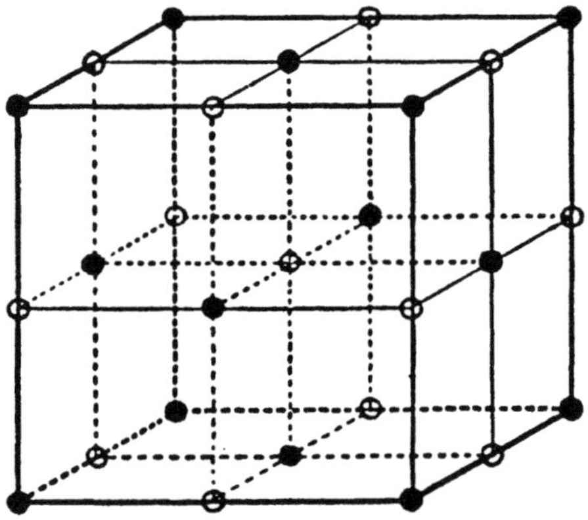

FIG. 28.—Structure of a rock-salt crystal.

In a *crystal of rock-salt,* the atoms form a *simple cubic lattice* in the sense of the considerations of the earlier sections (§§ 17 and 18), if we disregard the difference between the Na and Cl atoms. For the Na atoms by themselves form a

[1] *Cf.*, for instance, Sir W. H. and Prof. W. L. Bragg's "*X-Rays and Crystal Structure*" (London: Bell, 1924).

face-centred lattice, as do the Cl atoms likewise, and the two lattices are combined in such a way that the points of the one lattice lie at the mid-points of the edges of the other lattice (*cf.* Fig. 28, where the two kinds of atoms are distinguished by black and by white circles).

Now once the structure of rock-salt crystals is known, the *distance between the lattice planes* can immediately be calculated, taking the faces of the cubes as lattice planes. Denoting this distance by d, we see that each cube, the length of whose edges is d, contains a total of eight atoms, which are situated at its corners. Since, however, each corner belongs to the eight cubes which meet at it, the number of atoms contained in a cubic centimetre is given by

$$(1) \qquad Z = \frac{1}{d^3}.$$

On the other hand, if M be the so-called *molecular weight* of rock-salt, ϱ its *density*, and L the *Loschmidt number*,

$$(2) \qquad \varrho = \frac{M}{L} \cdot \frac{Z}{2}.$$

We have to divide by two, because it is the mass of *two* atoms (one Na and one Cl atom) taken together that is equal to the molecular weight divided by the Loschmidt number. Combining eqns. (1) and (2), we find the distance between the lattice planes to be given by [2]

$$(3) \qquad d = \sqrt[3]{\frac{M}{2 L \varrho}}.$$

The molecular weight of NaCl is equal to the sum of the atomic weights of Na (23·00) and Cl (35·46), *i.e.*,

$$(4) \qquad M = 58{\cdot}46.$$

The value found for the density of rock-salt from very accurate measurements is

$$(5) \qquad \varrho = 2{\cdot}164.$$

Hence, by eqn. (3) and § 88 eqn. (6),

$$(6) \qquad d = 2{\cdot}814 \times 10^{-8} \text{ cm}.$$

This value has been derived with an accuracy of about 1/3%, depending chiefly on the uncertainty occasioned by the Loschmidt number. Since, however, Röntgen-spectroscopic measurements can be carried out with a much greater accu-

[2] In other lattice structures the numerical factor in eqn. (3) would, of course, be different.

racy, an *ideal value* of $2{\cdot}81400 \times 10^{-8}$ cm. is assigned to d for such purposes.

As soon as a fundamental constant in the theory of crystals is thus evaluated, very exact *measurements of the wave-lengths* of Röntgen rays at once become possible. These may be most simply carried out by means of the *method of the rotating crystal*, devised by *W. H.* and *W. L. Bragg* in 1913. This method is based, on the one hand, on the law of reflection at the lattice planes derived in the previous section and likewise due to the Braggs, and, on the other hand, on the fact that, as far as the *reflection of Röntgen rays by crystals* is concerned, practically only the set of lattice planes *parallel to the front face of the crystal* requires consideration.[3]

Accordingly, a very narrow pencil of Röntgen rays from a small lead diaphragm is directed upon a crystal, which is mounted on a goniometer table, rotating about a vertical axis, in such a way that the axis of rotation lies in the front face of the crystal.[4] The crystal is very slowly rotated by hand, or by clockwork. According to § 18 eqn. (13), rays of the wave-length λ are then reflected only for quite definite glancing angles, *viz.*, only for those angles which, for integral (and low) values of the ordinal number n, satisfy the equation

$$2d \sin \theta = n\lambda. \tag{7}$$

As the crystal rotates, the ray glides along its face; the advantage of this method over one where the crystal is kept in the same position is that the disturbing effects of the unavoidable defects in the crystal face are thereby eliminated.

Now it follows from the geometrical considerations of § 18 that rays of the same wave-length are *focussed* at points on the circumference of a circle, whose centre lies on the axis of rotation, and which passes through the slit. If, therefore, a *photographic film* be placed round this circle, and if the radiation contains only rays of definite wave-lengths, *line spectra* of several orders must be formed on the film. The spectrum of the first order can at once be recognized, for it is far more intense than that of the second order, while the latter is more intense than the spectrum of the third order, and so on.

In general, only the *spectrum of the first order* is utilized. The wave-lengths of the lines contained in it are readily found by constructing a *scale* of wave-lengths in accordance

[3] Naturally, the face itself is always a lattice plane.

[4] In *Seeman's* apparatus a sharp edge is fixed directly against a part of the crystal face that is as perfect as possible. Edge and face together thus form a fine slit.

with eqn. (7) (wherein we put $n = 1$). The crystal of rock-salt may, of course, be replaced by any other, quite arbitrary crystal, even of unknown structure, provided only that one line can be identified with certainty, because of its origin always being the same.[5] A comparison of the positions of this line in the scales of the two crystals immediately enables the scale of the crystal used to be converted into that of rock-salt.

By means of the photographic method, a spectroscopic accuracy of measurement of almost one-millionth part of the measured value can be obtained. With this accuracy the wave-lengths of *different* lines can be compared, and, as was mentioned above, this necessitates the use of an ideal value of the rock-salt constant, which cannot be so accurately determined. The *photographic* method is due to *de Broglie* (1913); W. H. and W. L. Bragg were then still working with a so-called *ionization spectrometer*. This instrument consists of a lead cylinder filled with gas, and provided with a narrow slit through which Röntgen rays can pass into the gas. The cylinder is rotated around the crystal along the circle on which rays of equal wave-length are focussed, in such a way that the window always remains on the circle. Rays entering the window ionize the gas, and its resulting conductivity can be accurately measured. More recently the ionization method has been very much improved, and it is principally used for measurements of intensity in Röntgen spectra, for the intensity of the Röntgen rays is approximately proportional to the exactly measurable ionization. It may also be remarked that wave-lengths can be determined by the perfected ionization method with an accuracy of one in ten thousand.

If now we consider a small region of the spectrum extending from a wave-length λ to a wave-length $(\lambda + \Delta\lambda)$, the *spectral resolution* of this region is defined by the difference quotient $\Delta\theta/\Delta\lambda$, and hence approximately by the differential coefficient. It follows from eqn. (7) that

$$(8) \qquad \frac{d\theta}{d\lambda} = \frac{n}{2\,d\cos\theta}.$$

The resolving power thus increases with the ordinal number, but this advantage is counterbalanced by the rapid decrease in intensity with increasing ordinal number. Nevertheless, close doublets are often measured in the second order spectra, sometimes even in the third, fourth, or fifth orders. Moreover, as eqn. (8) shows, the resolution increases with the

[5] *Cf.* the remarks in § 21 on the subject of characteristic rays.

glancing angle θ, and it becomes especially strong when the angle θ approaches the value of 90°, so that sin θ is almost equal to unity. According to eqn. (7), however, this is the case when twice the distance between the lattice planes becomes nearly equal to an integral multiple of the wave-length of the region to be resolved. The resolution of a given small region, such as a doublet, is therefore best effected with a crystal which satisfies this relation for as small an ordinal number n as possible.

Since neither sin θ can be greater, nor n smaller than unity, it follows from eqn. (7) that the greatest wave-length which can ever be diffracted by a crystal is equal to twice the distance between the lattice planes, which in a rock-salt crystal is approximately equal to $5{\cdot}6 \times 10^{-8}$ cm.[6] The unit of wave-length used in Röntgen spectroscopy is the so-called *X-unit* (X.U.).[7] This is equal to one-thousandth of the optical spectroscopic unit, the Ångström unit, which, in its turn, is equal to the one-hundred millionth part of a centimetre. Hence

(9) $$1 \text{ X.U.} = 0{\cdot}001 \text{ Å.U.} = 10^{-11} \text{ cm.}$$

Thus with a rock-salt crystal, only wave-lengths up to about 5,000 X.U. can be measured. For the measurement of larger wave-lengths, other crystals with larger distances between the lattice planes are used, such as selenite or potassium ferrocyanide, in which the lattice plane distances are $7{\cdot}6 \times 10^{-8}$ and $8{\cdot}5 \times 10^{-8}$ cm. respectively. By means of these, wave-lengths as large as 13,000 X.U. can be measured. Using a crystal of palmitic acid, *Siegbahn* and *Thoraeus* have succeeded in measuring a wave-length of 17,660 X.U.[8] Apart from rock-salt, the principal crystal used in wave-length determinations is *calcite*, in which the distance between the lattice planes is $3{\cdot}029 \times 10^{-8}$ cm.[9] It may also be observed that soft Röntgen rays (*i.e.*, rays of large wave-length) are very strongly absorbed by gases even at quite low pressure, so that, in the determination of such wave-lengths, the whole path of the rays has to be *in vacuo*.

With very *hard* rays the method of the rotating crystal cannot be used. Hard rays penetrate too deeply into the crystal, and thereby produce a strong broadening effect on

[6] If the face of a regular crystal is not that of a cube, the lattice plane distance is less than d.

[7] The name is due to the fact that Röntgen rays are also called X-rays.

[8] *Arkiv. f. Math., Astr. och Fys*, **18** (1924), No. 24.

[9] Proposals have also been made to refer all wave-lengths to an idealized calcite constant ($d = 3{\cdot}02900 \times 10^{-8}$ cm.) instead of to the idealized rock-salt constant.

the lines. Hence, for very hard rays another method is used, which was developed by *Rutherford* and *Andrade*, in order to measure the wave-lengths of the extremely hard γ-rays. A widely diverging beam of rays falls on a thin plate of crystal, and the rays are reflected (at least, in a regular crystal) principally by the lattice planes that are normal to the crystal face struck by the rays. The crystal is fixed in one position, and behind it is placed a slit, which singles out from the emerging beam rays reflected at a certain glancing angle, and hence of a quite definite wave-length. By means of this method, into the details of which we cannot enter here, wave-lengths as small as 50 X.U. have been measured.

Finally, the last method of Röntgen ray analysis that we shall mention is the *crystal powder* method, developed on the one hand by *Debye* and *Scherrer*, and on the other hand by *Hull*. In this form, the time succession of diffraction centres in the rotating crystal method is replaced by a corresponding succession in space. The powder, in fact, consists of numerous minute crystals mingled together in all possible orientations. Hence, when a narrow pencil of rays strikes a small tube filled with the powder, reflection occurs at every possible angle, so that in this way also, the incident radiation is resolved into spectra.

§ 20. Duane and Hunt's Equation.

It follows from *Einstein's law of light-quanta* that, in the *generation of Röntgen rays by the arrest of cathode rays*, a single light-quantum must be produced from the kinetic energy of a single arrested electron. This kinetic energy is equal to the product of the elementary quantum of electricity (e) and the tube potential (V), the latter being given by the potential difference between cathode and anti-cathode.[1] Provided that the kinetic energy of the electron be *completely* transformed into a light-quantum, the following relation should hold :—

$$h\nu = eV, \tag{1}$$

where ν denotes the frequency of the Röntgen radiation generated. Hence, for the wave-length we have

$$\lambda = \frac{hc}{eV}. \tag{2}$$

If we measure in volts, the value of V must be multiplied by 300 ;[2] and if we express the wave-lengths in X-units,

[1] *Cf.* § 5.

[2] *Cf.* the prefatory note.

these must be multiplied by 10^{11}. Hence, giving the universal constants h, c, and e their known values, we find

$$(3) \qquad \lambda V = 1{\cdot}234 \times 10^7, \ (\lambda \text{ in X.U., } V \text{ in volts}).$$

This equation, however, only holds for the *limiting case* in which the kinetic energy of the arrested electron is *completely* transformed into a light-quantum and does not appear in any other form. In general, of course, this is not the case; for a portion of the kinetic energy is used up in other ways, and hence, in general, the *light-quantum* is *smaller* than the kinetic energy, whilst it can never be larger, according to the Law of the Conservation of Energy.

Thus, on the one hand, eqn. (3) determines the *largest possible frequency* or, what amounts to the same thing, the smallest possible wave-length, for a given tube potential; on the other hand, it determines the *lowest tube potential* necessary for the production of rays of a given wave-length *in the continuous Röntgen spectrum*. Eqn. (3) is generally known as *Duane* and *Hunt's* equation, after the investigators who first derived it (in 1915) from Einstein's law and confirmed it experimentally.

This equation forms an exact, quantitative expression for the observation made soon after the discovery of Röntgen rays, that the hardness of the rays increases with the tube potential. Later, more exact observations have shown that the *continuous Röntgen spectrum* exhibits a *sharp boundary towards the side of short wave-lengths*. According to eqn. (3), this boundary must be displaced in the direction of decreasing wave-length with increasing tube-potential, and eqn. (3) is therefore also known as Duane and Hunt's *displacement law*. Measurements carried out first by Duane and Hunt, and later by other investigators, have proved eqn. (3) to be satisfied with the greatest accuracy, as regards the limiting wave-length; the measurements covered a range from about 4,500 to 170,000 volts.[3] Indeed, the accuracy is so great that a determination of the elementary quantum of action can be based upon this displacement (assuming the value of e to be known).[4]

[3] In accordance with eqn. (3), the corresponding limiting wave-lengths lie between about 80 and 3,000 X.U.

[4] *Cf.* the comprehensive report by E. Wagner in the *Jahrb. d. Radioakt. u. Elektronik*, 16 (1920), p. 190.—It should, however, be borne in mind that the value of the Loschmidt number is also implicitly contained in the wave-length (in virtue of § 19 eqn. (3)]. Since the wave-length is inversely proportional to the cube root of the Loschmidt number, while the latter is equal to the quotient of Faraday's constant and the elementary quantum of electricity, it is not really the quotient e/h that occurs in eqn. (2), but the quotient $e^{\frac{4}{3}}/h$.

Eqn. (3) determines, as already explained, the minimum potential necessary for the generation of a given wave-length, or the so-called *excitation potential*. This may be experimentally determined with great accuracy. If the tube potential be gradually increased, the given wave-length is found to appear, though very weakly, at a quite definite potential, and, as soon as that minimum value is exceeded, the intensity of the radiation in the region of that wave-length is found to increase *linearly* with the potential. The intensity is determined by means of an ionization chamber (*cf.* § 19), the conductivity of the ionized gas being measured by the deflections of an electrometer. In Fig. 29, which reproduces the results of experiments made by Duane and Hunt in 1915, the dependence of the measured values of the ionization upon the potential, for various wave-lengths, is represented by black dots. As the figure shows, almost straight lines may be drawn through the dots, the intersections of these lines with the axis of abscissæ giving the excitation potentials for the wave-lengths in question. In this way, from each of the curves, Duane and Hunt were able to *calculate*, according to eqn. (2), the *elementary quantum of action.*[5] This method of the direct determination of h has subsequently been greatly improved. For example, *Duane* and *Blake*[6] obtained in later measurements a value of $(6{\cdot}557 \pm 0{\cdot}013) \times 10^{-27}$ erg sec., while the value derived by other methods, and accepted at present as the most correct, is $(6{\cdot}545 \pm 0{\cdot}012) \times 10^{-27}$ erg sec.

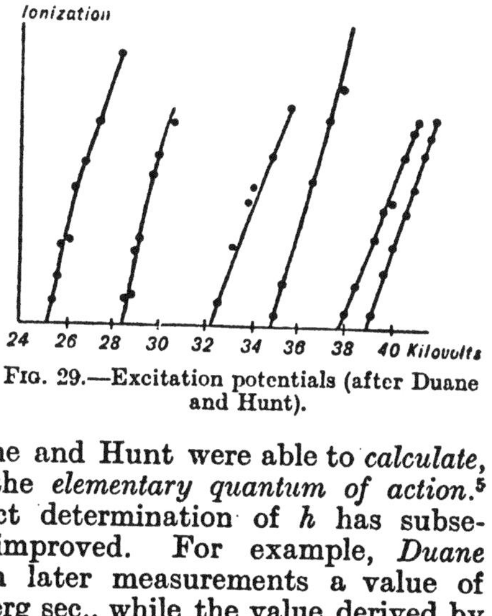

FIG. 29.—Excitation potentials (after Duane and Hunt).

Duane and Hunt's measurements thus represent a direct experimental confirmation of the quantum hypothesis. It must, however, be remembered that the relations based on the quantum hypothesis refer *only to the frequency* of the excited Röntgen radiation, but make no predictions on the

[5] From left to right, the six lines in Fig. 29 correspond to the wave-lengths 488, 424, 377, 345, 318, and 308 X.U. The values of h calculated from the six lines are: 6·44, 6·37, 6·34, 6·38, 6·41 and $6{\cdot}39 \times 10^{-27}$ erg sec.

[6] *Phys. Rev.*, 10 (1917), pp. 93 and 624.

subject of the energy relations. It is by no means implied that the kinetic energy of each and every arrested electron is transformed into a light-quantum. As is shown by measurements on the heat generated by Röntgen rays, only about *two thousandths of the energy* of the cathode rays in a Röntgen tube are transformed into energy of Röntgen radiation. In the few cases, however, in which a corpuscular quantum of energy is changed into a light-quantum, Einstein's relation must be obeyed, as, indeed, is proved by Duane and Hunt's measurements.

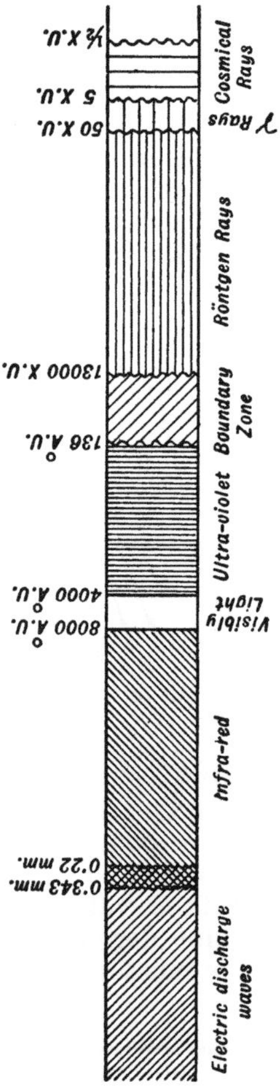

FIG. 30.—Complete spectrum.

Just like ultra-violet rays, Röntgen rays may, of course, also give rise to the so-called *photo-electric effect*, consisting in the *liberation of electrons* from the surface of irradiated bodies (*cf.* § 5). In consequence of their magnetic deflectability, these electrons can be isolated from the secondary Röntgen radiation which is emitted by a body subjected to the action of primary Röntgen rays. They were discovered by *Dorn* in 1900, in the form of so-called *secondary cathode rays*. Owing to the much greater frequency of the generating light, their velocity is much larger than that of the electrons liberated by ultra-violet light. In the case of the secondary cathode rays, Einstein's law of light-quanta is again found to be accurately obeyed; the kinetic energy of a liberated electron is equal to a light-quantum of the generating radiation, less the work expended in the liberation.

In 1923 it was discovered by *Compton* that the generation of secondary, continuous, Röntgen radiation by a primary radiation is accompanied by an increase in the wave-length. A quantitative interpretation of the observed effect is afforded on the assumption that the radiation is propagated in light-quanta, which can collide with free, or approximately free, electrons in such a way that both the Law of the Conservation of Energy and the Law of the Conservation of the Total Momentum are satisfied. The liberation of energy from a light-quantum must thus lead to a diminution in its frequency. In Compton's experiments, the primary radiation emitted by a molybdenum anticathode was scattered by charcoal or paraffin wax; the loosely bound valency electrons of the carbon then behave as though they were free. The secondary spectrum appears displaced in the direction of longer wave-lengths with respect to the primary spectrum. In agreement with the experimental results, the theoretical relation expressing the (positive) displacement of the wave-length is given by

$$\Delta\lambda = 2\lambda_0 \sin^2 \frac{\theta}{2}. \tag{4}$$

In this expression θ is the angle of scattering, and $\lambda_0 = h/mc = 0{\cdot}024$ Å is the wave-length of a light-quantum, the mass of which (calculated according to the Law of the Inertial Mass of Energy) is equal to the mass of an electron.

The quantum relations holding for Röntgen rays also enable an *indirect determination of Röntgen wave-lengths* to be made, which is of special importance in those regions of the spectrum where a direct determination by means of crystals is not possible. Thus, on the ground of Duane and Hunt's equations, very large wave-lengths up to 40,000 X.U. could be determined by measuring the tube potential.[7] Furthermore, the equation of the photo-electric effect renders possible an indirect determination of wave-lengths, as will later be discussed at length. In virtue of this effect it has been possible to measure, on the one hand, long waves up to about 40,000 X.U. in length,[8] and, on the other hand (as we shall see in detail later), wave-lengths as short as 20 X.U. in the region of the γ-rays.

Of still shorter wave-lengths than the γ-rays are the rays of the so-called *penetrating radiation* detected in the upper regions of the atmosphere by *Hess* and *Kohlhörster* in 1912–1914. The wave-lengths of these rays, which are evidently of cosmic origin, were determined in 1925 by *Millikan.*[9] He found the extraordinarily small value of about 0·18/metre for the absorption coefficient in water, and wave-lengths as small as 1/2 X.U. The rays are not homogeneous, and are possibly secondary rays which, in

[7] *Cf.* H. M. Dadourian, *Phys. Rev.*, **14** (1919), p. 234.
[8] *Cf.* § 21.
[9] *Nature*, **116** (1925), p. 823.

accordance with the Compton effect, are not so hard as the primary cosmic rays entering the atmosphere from space.

Thus, whereas the Röntgen rays measured directly by means of *crystals* extend over a region of some *eight octaves* (from about 50 to about 13,000 X.U.), the above-mentioned indirect methods have resulted in the addition of approximately seven octaves on the side of shorter, and of approximately two octaves on the side of longer wave-lengths. About one octave above the longest wave-length in the region of the Röntgen rays (40,000 X.U.) lies the shortest wave-length hitherto measured in the ultra-violet—136 Å.U. (= 136,000 X.U.). We can, however, scarcely speak of a true gap, because (as we shall discuss more fully later) the atomic spectra of the elements establish a complete connection between ultra-violet rays and Röntgen rays.

The whole spectrum of electromagnetic waves is represented in Fig. 30. The unit of scale is an octave, so that, as regards wave-lengths, the scale is logarithmic. From the largest wave-length measured in the infra-red (0·343 mm.) down to the smallest known wave-length (*c.* 1/2 X.U.), the spectrum extends over some *thirty-one octaves.* Of these about eight lie in the infra-red, about five in the ultra-violet, and about sixteen are covered by Röntgen rays (in the wider sense), whilst the whole of the rays visible to the human eye are comprised in the narrow limits of a single octave.[10]

[10] The shortest electric discharge wave yet obtained (0·22 mm.) is that measured by Nichols and Tear in 1923. [*Proc. Nat. Acad.*, Washington, **9** (1923), pp. 211-214.] Until that date, the shortest known electric discharge wave was one of about 3 mm. length produced by Lampa in 1897. The longest known infra-red wave, 0·343 mm. in length, was discovered in 1910 by Rubens and Bayer in the radiation of great wave-length emanating from a mercury lamp.

CHAPTER IV

THEORY OF THE ELEMENTS

§ 21. Moseley's Law, and the Natural Sequence of the Elements.

IN 1905 *Barkla* made the important discovery of *characteristic Röntgen radiation.* As Barkla observed, when primary Röntgen rays impinge on a body, the *secondary* Röntgen radiation emitted by it consists chiefly of *homogeneous rays,* the *hardness* of which is found to be a *characteristic property of the atoms* of the body. Barkla measured the hardness of the rays by means of their *absorption* in aluminium plates, the absorption coefficient representing a constant characteristic of the element in question.[1] He found the radiation from chemical compounds to be composed of the rays of the elements contained in the compounds. He further observed that the hardness of the characteristic rays *increases with the atomic weight,* and, moreover, that the majority of the elements appeared to possess not merely one, but *two* absorption coefficients differing widely in value. Barkla therefore distinguished between a harder *K-radiation* and a softer *L-radiation.* He attributed quite rightly to the limitations of experimental methods the fact that only one species of rays could be demonstrated in the case of many elements. For, on the one hand, the *K*-rays become so hard with increasing atomic weight that, in the state of Röntgen technique at that time, it was impossible to excite them; on the other hand, the *L*-radiation becomes so soft with decreasing atomic weight that it could not be demonstrated by the methods then known. Furthermore, as Barkla discovered, characteristic rays are generated only by primary rays which are *harder* than the characteristic rays themselves.[2]

[1] The reciprocal value of the absorption coefficient represents the depth (in centimetres) to which the rays penetrate in aluminium before their intensity is diminished in the ratio $e : 1$.

[2] This law is analogous to the well-known Stokes's rule, according to which the wave-lengths of fluorescent light are greater than those of the exciting light. On that account, the characteristic radiation was also called by Barkla fluorescent radiation.

The invention of the method of the *rotating crystal* rendered possible an exact analysis of the characteristic radiation, and led to its resolution into a *line spectrum*, the exact measurement of *wave-lengths* taking the place of the inaccurate determination of hardness. We owe the foundation of *Röntgen spectroscopy* to *Moseley* who, in 1913, by means of a specially constructed vacuum spectrograph, resolved the K- and the L-radiations of numerous elements into distinct sharp lines, whose wave-lengths he determined accurately. Moseley found that both the K- and the L-series are of the same type for the different elements. In each series one line was found to be particularly strong, and this line he called the K_α- and the L_α- line respectively. More recent Röntgen-spectroscopic researches (which we shall discuss more fully in a later section) have resulted both in the observation of new lines in both series, and also in the discovery of two further series. These, on account of their softness, have actually been demonstrated only in elements of high and the highest atomic weights, and are known as the M- and the N-series.[3]

The increase in hardness with the atomic weight had been established by Barkla, but Moseley succeeded in finding a much more exact expression for this relation in a law which, discovered by him, was destined to be of fundamental importance for the modern *theory of the elements*. As Moseley discovered, the *elements* can be arranged in a *consecutively numbered series*, in such a way that the *square root of the frequency of a definite Röntgen line increases linearly with the number of the element.*

The "*natural*" *sequence of the elements* thus obtained agrees on the whole, as Moseley recognized, with the series formed by arranging the elements in order of *increasing atomic weight.* Whereas, however, no regular relation exists between the atomic weight numbers, in the natural sequence the displacement of the Röntgen lines from element to element takes place with the greatest exactness, so that the existence of any gap in the series is clearly revealed by too large a jump.

In Fig. 31, some photographs taken by Moseley of the K- series for a sequence of elements are reproduced. In addition to the K_α- line, a second, weaker, (K_β-) line can be recognized. (It was not until spectroscopic methods had been perfected that further lines were first observed.) The

[3] The N-series is of longer wave-length than the M-series and the wave-lengths of the M-series are greater than those of the L-series. *Cf.* § 30.

photographs are so arranged that points lying vertically with respect to each other correspond to the same wave-length. The wave-lengths increase from left to right, and the frequencies from right to left, so that, in Fig. 31, the element with the lower number lies above the element with the higher number. The photographs refer to the elements calcium, titanium, vanadium, chromium, manganese, iron, cobalt, nickel, copper, and zinc. Zinc was used alloyed with copper, in the form of brass.[4] The exposures were made with interchangeable anti-cathodes of the elements to be investigated, so that the characteristic radiation was directly generated by the cathode rays.[5] The photographs show with wonderful clearness that the given elements follow the natural sequence, but that there is an element missing between calcium and titanium. (This is scandium, which at the time was not available to Moseley, because of its rarity.) It may also be noticed—and this is most remarkable—that cobalt precedes nickel in the natural sequence, although it has the higher atomic weight of 58·97 as against 58·68. Brass shows both the zinc and the copper lines, and cobalt appears to be contaminated by nickel and iron.

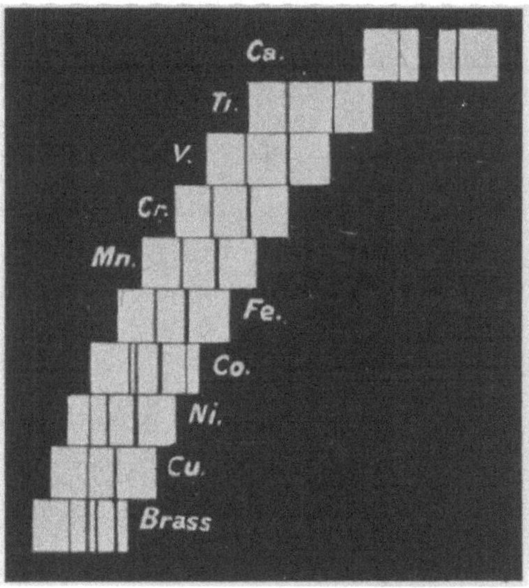

FIG. 31.—*K*-series (after Moseley).

Direct Röntgen-spectroscopic measurements, carried out with the help of crystals (as will be more fully described in a later section) have made it possible to establish the natural

[4] The alloy was used because zinc is not sufficiently resistant to heat

[5] Moseley utilized a slit near the anti-cathode, so that the emerging pencil of rays was widely divergent. In this way he was able to photograph both *K*-lines in one exposure, without rotating the crystal, provided the latter was suitably placed. The subsequent introduction of the method of the rotating crystal resulted in a considerable improvement in the photographs.

sequence of the elements from sodium to the highest element *uranium*. Furthermore, indirect measurements of characteristic wave-lengths in the boundary region between Röntgen rays and ultra-violet rays have enabled us (as we shall discuss more fully later) to follow the course of the spectra from sodium down to the element with the second smallest atomic weight, helium.

The natural sequence, which undoubtedly commences with hydrogen in the first and helium in the second place, is thus *unequivocally* established. Including uranium, it comprises *ninety-two places*, only *three* of which belong to *unknown elements* (Nos. 61, 85, 87). The natural sequence of the elements obtained in this manner is represented by Table III, in which, for each element, the *atomic number* or *ordinal number*, which determines its place in the natural sequence, the conventional name, the chemical symbol, and the atomic weight [6] relative to oxygen = 16·000 are given. This table, it is true, includes elements for which it has not been possible to make Röntgen-spectroscopic photographs. These elements comprise the rare gases, mercury (which could not be applied as an anti-cathode material either in the metallic form or in compounds), and some particularly rare elements (thulium, radium, polonium, actinium, protoactinium). Apart from the chemical periodicity, the positions of these elements in the natural sequence follow from their atomic weights.

Table III

The Natural Sequence of the Elements

1	Hydrogen . .	H	1·008	17	Chlorine . . .	Cl	35·46
2	Helium . . .	He	4·00	18	Argon . . .	A	39·91
3	Lithium . . .	Li	6·94	19	Potassium . .	K	39·10
4	Beryllium . .	Be	9·02	20	Calcium . . .	Ca	40·07
5	Boron . . .	B	10·82	21	Scandium . .	Sc	45·10
6	Carbon . . .	C	12·00	22	Titanium . .	Ti	48·1
7	Nitrogen . . .	N	14·008	23	Vanadium . .	V	51·0
8	Oxygen . . .	O	16·000	24	Chromium . .	Cr	52·01
9	Fluorine . . .	F	19·00	25	Manganese . .	Mn	54·93
10	Neon	Ne	20·2	26	Iron	Fe	55·84
11	Sodium . . .	Na	23·00	27	Cobalt . . .	Co	58·97
12	Magnesium . .	Mg	24·32	28	Nickel . . .	Ni	58·69
13	Aluminium . .	Al	26·97	29	Copper . . .	Cu	63·57
14	Silicon . . .	Si	28·06	30	Zinc	Zn	65·37
15	Phosphorus . .	P	31·03	31	Gallium . . .	Ga	69·72
16	Sulphur . . .	S	32·06	32	Germanium . .	Ge	72·60

[6] The atomic weights are those given in the report of the "International" Committee on Chemical Elements for 1925. The names and symbols allotted to elements Nos. 43 and 75 are provisional, and are not yet generally accepted.

TABLE III—*continued.*

33 Arsenic	As	74·96
34 Selenium	Se	79·2
35 Bromine	Br	79·92
36 Krypton	Kr	82·9
37 Rubidium	Rb	85·44
38 Strontium	Sr	87·6
39 Yttrium	Y	89·0
40 Zirconium	Zr	91
41 Columbium	Cb	93·1
42 Molybdenum	Mo	96·0
43 Masurium	Ms	—
44 Ruthenium	Ru	101·7
45 Rhodium	Rh	102·9
46 Palladium	Pd	106·7
47 Silver	Ag	107·88
48 Cadmium	Cd	112·41
49 Indium	In	114·8
50 Tin	Sn	118·7
51 Antimony	Sb	121·8
52 Tellurium	Te	127·5
53 Iodine	I	126·93
54 Xenon	Xe	130·2
55 Cæsium	Cs	132·8
56 Barium	Ba	137·4
57 Lanthanum	La	138·90
58 Cerium	Ce	140·2
59 Praseodymium	Pr	140·9
60 Neodymium	Nd	144·3
61 —	—	—
62 Samarium	Sm	150·4
63 Europium	Eu	152·0
64 Gadolinium	Gd	157·3
65 Terbium	Tb	159·2
66 Dysprosium	Dy	162·5
67 Holmium	Ho	163·4
68 Erbium	Er	167·7
69 Thulium	Tm	169·4
70 Ytterbium	Yb	173·6
71 Lutecium	Lu	175·0
72 Hafnium	Hf	178·3
73 Tantalum	Ta	181·5
74 Tungsten	W	184·0
75 Rhenium	Re	—
76 Osmium	Os	190·8
77 Iridium	Ir	193·1
78 Platinum	Pt	195·2
79 Gold	Au	197·2
80 Mercury	Hg	200·6
81 Thallium	Tl	204·4
82 Lead	Pb	207·2
83 Bismuth	Bi	209·0
84 Polonium	Po	(210)
85 —	—	—
86 Radium emanation	Ra Em	222
87 —	—	—
88 Radium	Ra	226·0
89 Actinium	Ac	(226)
90 Thorium	Th	232·1
91 Protoactinium	Pa	(230)
92 Uranium	U	238·2

In the natural sequence, the atomic weight increases throughout, except in four instances. The heavier element argon precedes the lighter element potassium (Nos. 18 and 19); cobalt comes before nickel (Nos. 27 and 28) as already mentioned; tellurium occurs before iodine (Nos. 52 and 53); and, finally, thorium precedes protoactinium (Nos. 90 and 91).

Now, for hydrogen and helium, the atomic number agrees with the nuclear charge number, as follows from our considerations of the line spectra of hydrogen and ionized helium. It therefore seems plausible to *identify* quite generally the *atomic numbers* of the elements with their *nuclear charge numbers*, which, referred to the elementary quantum as unit, give the positive charge of the atomic nucleus, and at the same time the *number of electrons revolving around the nucleus* in the *neutral* atom. The strongest support for this conception, proposed by *van den Broek* in 1913, is to be found in Bohr's general formula for the frequency of a spectral line in a hydrogen-analogue [§ 8 eqn. (2)]. For, according to this formula, the square root of the frequency of a spectral line due to a definite transition increases linearly with the *nuclear charge number*, while, in accordance with Moseley's law, it increases linearly with the *atomic number*. Hence, if the atomic number agrees with the nuclear charge number for *one* element (as with helium), the agreement must obviously hold for *all* the elements. The correctness of van den Broek's assumption is also proved by the numerical relations of chemical periodicity which we shall discuss in a later section.

Moreover, this hypothesis has been directly confirmed by experimental researches on the *passage of α-rays through metal foils*. As we have already mentioned in an earlier section (§ 4), α-rays are deflected by the atomic nucleus in their passage through matter. The so-called *scattering* of the α-rays thus produced therefore depends on the nuclear charge number of the metal traversed. From measurements carried out on the scattering produced by metal foils *Chadwick*, in 1920, was able in this manner to calculate the nuclear charge number of three metals with an accuracy of 1%. For platinum, silver, and copper, Chadwick found the values 77·4, 46·3, and 29·3. Within the limits of error of 1%, these values do, indeed, agree very well with the atomic numbers 78, 47, and 29, given in Table III.

§ 22. The Periodic System.

In the natural sequence of the elements, experience shows that *six elements* occupy a *very special position*, while simple numerical relations exist between their atomic numbers. These elements are, in fact, absolutely *passive* from the *chemical* point of view ; in contrast to all the other elements, they form *no chemical compounds* whatsoever. They are the so-called *rare gases*,[1] *viz.*, helium, neon, argon, krypton, xenon, and radium emanation.[2] Their atomic numbers are

2, 10, 18, 36, 54, 86.

The differences between these numbers are

8, 8, 18, 18, 32,

[1] The rare gases (apart from radium emanation) were discovered by *Ramsay* in the atmosphere, towards the end of the nineteenth century.

[2] The "International" Committee on Chemical Elements has proposed the names *radon, thoron*, and *actinon* for the three radioactive emanations. *Radon* thus replaces *niton*, but the term *radium emanation* is also still widely used.

and we may suspect in advance a fundamental reason for the fact that the first atomic number and the resulting differences between the successive numbers are of the form $2z^2$ ($z = 1, 2, 3, 4$).

From the point of view of modern atomic theory, *chemical activity* must obviously be referred to *electrical forces* proceeding from the atoms. It is these electrostatic forces which lead to a combination of oppositely charged atoms, and thus to the *formation of molecules*. Since, in the neutral atom, the number of electrons revolving around the nucleus is equal to the nuclear charge number, an atom can only become chemically active when the number of electrons surrounding its nucleus becomes either more or less than the nuclear charge number. In the first case, the atom as a whole becomes electrically negative; in the second case, it becomes positive; moreover, the charge is determined by the number of electrons gained or lost. It must therefore be concluded from the chemical passivity of the rare gases that, in their atoms, the *arrangement of the electrons about the nucleus* is generally *stable*, in the sense that electrons are neither removed from the atomic system, nor are new ones introduced into it—a state of affairs that appears possible in all other atoms.

Thus the rare gases to a certain extent represent landmarks in the natural sequence of the elements, and divide the series into *seven periods*. Of these, the first period contains the first *two* elements (hydrogen and helium); the second and third periods each contain *eight* elements, Nos. 3 to 10 (lithium to neon), and Nos. 11 to 18 (sodium to argon). The fourth and fifth periods each comprise *eighteen* elements, Nos. 19 to 36 (potassium to krypton), and Nos. 37 to 54 (rubidium to xenon). The sixth period includes *thirty-two* elements, Nos. 55 to 86 (cæsium to radium emanation). The seventh period, beginning with the unknown element No. 87, apparently breaks off in the sixth place at uranium; and it must remain an open question whether elements with still higher nuclear charge are not capable of existence, have already disintegrated, or have not yet been discovered because of their rarity.[3]

As long ago as 1869, *Lothar Meyer* and *Mendeléeff* simultaneously, but independently, made the important discovery that, if the elements be arranged in order of increasing atomic weight, characteristic physical and chemical properties recur *periodically*, and that, if the successive periods be written

[3] *Cf.* § 26.

TABLE IV

The Periodic System of the Elements

	I		II		III		IV		V		VI		VII		VIII	
	A	B	A	B	A	B	A	B	A	B	A	B	A	B	A	B
1	1 H															2 He
2	3 Li		4 Be			5 B		6 C		7 N		8 O		9 F		10 Ne
3	11 Na		12 Mg			13 Al		14 Si		15 P		16 S		17 Cl		18 A
4	19 K	29 Cu	20 Ca	30 Zn	21 Sc	31 Ga	22 Ti	32 Ge	23 V	33 As	24 Cr	34 Se	25 Mn	35 Br	26 Fe 27 Co 28 Ni	36 Kr
5	37 Rb	47 Ag	38 Sr	48 Cd	39 Y	49 In	40 Zr	50 Sn	41 Cb	51 Sb	42 Mo	52 Te	43 Ms	53 I	44 Ru 45 Rh 46 Pd	54 Xe
6	55 Cs	79 Au	56 Ba	80 Hg	57—71 rare earths	81 Tl	72 Hf	82 Pb	73 Ta	83 Bi	74 W	84 Po	75 Re	85	76 Os 77 Ir 78 Pt	86 RaEm
7	87		88 Ra		89 Ac		90 Th		91 Pa		92 U					

down in successive horizontal rows, an arrangement is possible in which elements occurring in the same vertical column are distinguished by similar properties. The further development of the so-called *periodic system* resulted in a distinction between *eight vertical groups*, each with two sub-groups. An exact representation, however, was not possible until the knowledge of atomic numbers also led to an exact recognition of the number of places in the periods.

Experience shows that the table of the periodic system is to be constructed in the following manner (Table IV.). Eight vertical columns are formed, each with two sub-groups. The groups corresponding to the vertical columns are denoted by I, II, *etc.*, up to VIII; the sub-groups may be distinguished in the table by the letters A and B. The two elements in the first period are now placed in sub-groups I A and VIII B. The eight elements each of the second and the third period are to be allotted to the eight groups in such a way that the first two in each period fall in A columns, and the last six in B columns. In the fourth and fifth periods, each containing 18 elements, we proceed as follows: we place the first eight elements in the eight A sub-groups, and the last eight in the eight B sub-groups, whilst the ninth and tenth elements are allotted to sub-group VIII A, so that, in each of the fourth and fifth periods, this sub-group contains a *triad*. In the sixth period we follow the same arrangement as in the fourth and fifth, but with the difference that in sub-group III A a *group of fifteen elements* is to be placed, instead of a single element. Thus, in the sixth period, one of the 16 positions in the eight A and eight B sub-groups will contain 3 elements, and another 15 elements. Finally, in the seventh period, the six elements are to be placed in the A sub-group.

The three triads consist of the *iron*, the *palladium*, and the *platinum metals*. The group of fifteen elements is known as the *rare earth* group. Furthermore, the periodic system indicates to us the chemical nature of elements as yet unknown. Thus, in 1925, the two missing *homologues of manganese*, Nos. 43 and 75, were discovered by *Noddack* and *Tacke*[4]. Of the three elements not yet known, the 61st must be a rare earth, the 85th a halogen, and the 87th an alkali metal.

§ 23. Isotopy.

As we have seen in the previous considerations, the position of an element in the periodic system depends solely on its *nuclear charge*. Furthermore, we must assume

[4] *Sitz. Ber. Berlin. Akad.*, June 11th, 1925.

the *nuclei* of the elements to be *built up* of protons and electrons. Hence, provided only that the *resultant nuclear charge*—the *algebraical sum* of all the positive and negative elementary charges forming the nucleus—be the same in two atoms, these must exhibit the same chemical behaviour, even though the number of protons and nuclear electrons differs in the two. Whereas, however, the nuclear charge, and thus the *chemical character*, appears to be determined by the *difference* in the numbers of protons and electrons in the nucleus, the *mass* of the atom, and therefore the so-called *atomic weight*, depends *only* on the number of *protons*, since the mass of an electron can be neglected in comparison with that of a proton. It is thus quite possible for two substances, despite *different atomic weights* and *different nuclear structures*, to agree in their chemical and in many important physical properties to such an extent that they would really represent only *different species of one and the same element*.

The first phenomenon of this kind to attract the notice of physicists was the *inseparability* of certain *radioactive elements*. When, for instance, salts of *thorium* and *ionium* (the parent substance of radium) were in any way mixed together, it was found quite impossible to separate them again by any method whatsoever; moreover, the *spectra* of thorium and ionium were observed to be quite identical. Similarly, radium and mesothorium 1 were found to be inseparable, as also were lead and radium D (*cf.* § 26).

All these cases proved that substances differing in their radioactive behaviour could yet be identical in chemical and also in physical respects. Recognition of this fact led *Soddy*, in 1910, to the belief that also the *inactive* (non-radioactive) *elements* might represent *mixtures of several, chemically inseparable species*. Soddy further pointed out that this assumption would solve a great difficulty which still faced atomic theory at that time. This difficulty arose from the *deviations* of the so-called *atomic weights from whole numbers*.

If atoms are actually built up of protons (*i.e.*, hydrogen nuclei and electrons), and if the mass of the electrons can be neglected in comparison with that of the protons, then all atomic weight numbers must obviously be *integral*. Now, although this is the case with many elements, and especially with the most abundant ones, yet in other elements such as chlorine, the deviations from whole-number values are so considerable that they could not at first be reconciled with that necessary consequence of atomic theory. This difficulty, however, vanishes at once if, according to Soddy's assumption, the elements really represent *mixed elements.* For, in that case, only the significance of an average weight would attach to the atomic weight, and its constancy would be explained on the grounds that any mixture ratio once formed (*e.g.*, in times when the earth was in a gaseous state), must remain unaltered for ever, in view of the inseparability of the mixed atoms.

This conception, suggested by Soddy in 1910, was experimentally confirmed in 1913 by *J. J. Thomson*, who showed by means of *positive ray analysis* that the rare gas *neon* must really contain *two atomic species* of *different atomic weights* (20 and 22). Accurate experiments made in 1914 also confirmed the opinion, expressed by Soddy, that the *atomic weight of lead* must differ in value, according as to whether we are dealing with "ordinary" lead or with lead which has been obtained from uranium minerals or from thorium minerals. (As we shall show in § 26, both uranium and thorium are gradually transformed into lead.) Whereas the atomic weight of ordinary lead is found to be 207·2, atomic weights as low as 206·05 have actually been determined for lead from uranium minerals, and as high as 207·9 for lead from thorium minerals.[1] The term *isotopes*, suggested by Soddy, soon came into general use to denote substances which have different atomic weights, but exhibit the same chemical properties.[2]

Researches on isotopes, properly speaking, began in 1919 with *Aston's* development of positive ray analysis into the so-called *mass-spectroscopy*, by means of an artifice already described (*cf.* § 4). Using his new method, Aston was first able to prove, with greater accuracy, that the rare gas *neon* does actually consist of *two isotopes* with the exactly *integral atomic weights* 20·0 and 22·0. Shortly afterwards, he was also able to prove that *chlorine*, in which the deviation from a whole-number value had always been felt to be particularly embarrassing, is a mixture of two isotopes with the exactly integral atomic weights 35·0 and 37·0. It follows from the mixture weight (35·46) that the ratio of the lighter to the heavier gas is about three to one, as can also be recognised from the intensity of the lines in the mass-spectra.

[1] *Cf.* Hönigschmid, *Zeitschr. f. Electrochemie*, **24** (1918), p. 163, and **25** (1919), p. 91. *Cf.* also § 26, note 7.

[2] *Isos topos* in Greek means "the same place." The name was suggested by the fact that isotopes occupy the same place in the periodic system of the elements.

By means of his mass-spectrograph, Aston has investigated numerous other elements for isotopes. This method, however, has necessarily been limited to such elements as can be introduced, in a gaseous state at ordinary temperature into a Geissler tube, whether as element or in the form of a chemical compound. Subsequently, the analysis of non-volatile substances has also been successfully carried out, by introducing them on the *anode* of the discharge tube and investigating the anode or positive rays produced. Using this method, *Aston*, *G. P. Thomson*, and *Dempster* have made further analyses.[3]

In Fig. 32 (Plate II), Aston's *mass-spectra* of neon, chlorine, argon, and krypton are reproduced, showing most clearly the integral nature of the pure atomic weights.[4] Aston's method is sufficiently delicate to be able to detect an isotope present only to the extent of 1 in 1,000 in a mixture. The degree of blackening of the individual lines enables quantitative conclusions to be drawn on the mixture ratios of the isotopes; and it is possible in this way to determine the mixture weight by a mass-spectroscopic method[5] (though Dempster's method gives more accurate results).

In Table V are reproduced the results of researches (up to the end of 1925) on the subject of isotopes in the non-radioactive elements; it contains 56 elements and 120—130

[3] In Dempster's method the anode rays, which were deflected only by a magnetic field, were caught on a metal plate connected to an electrometer. In this way the relative quantities of the isotopes in the mixed element could be fairly accurately estimated.

[4] Lines of known atoms or groups of atoms serve as reference lines in these spectrograms, especially the lines of carbon (12); the C-complexes, CH, CH_2, CH_3, CH_4 (13, 14, 15, 16); the C_2-complexes (24, 25, 26, 27, 28, 29, 30); O (16), OH (17), OH_2 (18), CO (28), and CO_2 (44). Spectrum I shows, besides these lines, which almost always occur, the two neon lines very clearly at 20 and 22. In spectra II, III, and IV can be seen the lines of Cl_{35} and Cl_{37}, as well as lines at 36 and 38 due to $Cl_{35}H$ and $Cl_{37}H$. We may also notice "second order lines" at 17·5 and 18·5. These are produced by chlorine ions with double charges, so that the specific charge is then twice as large, which would also be the case if the charge remained single and the mass were halved. Lines 63 and 65 are due to $COCl_{35}$ and $COCl_{37}$. Spectra V and VI show not only the first order lines of argon, but also, quite clearly, a second order line at 20 and a third order line at 13·3. Spectrum VIII is that of krypton, showing plainly five strong lines at 80, 82, 83, 84, 86 and a sixth, weaker, line at 78. These lines also occur in the second order in the mass-spectrum of krypton. An explanation of spectrum VII may well be omitted here. For details of Aston's mass-spectroscopy, *vide* his book, "*Isotopes*" (London: Arnold, 1924).

[5] In the case of boron, it was actually recognized on these grounds that its "atomic weight" must be lower than that previously accepted. Accurate, direct chemical determinations of the atomic weight later proved that the atomic weight of boron, as stated, was really one unit too high in the first decimal place.

TABLE V

Isotopes

Atomic Number.	Element.	Atomic Weight.	No. of Isotopes.	Mass-numbers of Isotopes in order of relative amounts in mixture.
1	Hydrogen .	1·008	1	1·008
2	Helium . .	4·00	1	4
3	Lithium .	6·94	2	7, 6
4	Beryllium .	9·02	1	9
5	Boron . .	10·82	2	11, 10
6	Carbon . .	12·00	1	12
7	Nitrogen .	14·008	1	14
8	Oxygen . .	16·000	1	16
9	Fluorine .	19·00	1	19
10	Neon . .	20·2	2 (3)	20, 22, 21
11	Sodium . .	23·00	1	23
12	Magnesium .	24·32	3	24, 25, 26
13	Aluminium .	26·97	1	27
14	Silicon . .	28·06	3	28, 29, 30
15	Phosphorus .	31·03	1	31
16	Sulphur .	32·06	1	32
17	Chlorine .	35·46	2	35, 37
18	Argon . .	39·91	2	40, 36
19	Potassium .	39·10	2	39, 41
20	Calcium .	40·07	2	40, 44
21	Scandium .	45·10	1	45
22	Titanium .	48·1	1 (2)	48, (50)
23	Vanadium .	51·0	1	51
24	Chromium .	52·01	1	52
25	Manganese .	54·93	1	55
26	Iron . .	55·84	2	56, 54
27	Cobalt . .	58·97	1	59
28	Nickel . .	58·69	2	58, 60
29	Copper . .	63·57	2	63, 65
30	Zinc . .	65·37	4	64, 66, 68, 70
31	Gallium .	69·72	2	69, 71
32	Germanium .	72·60	3	74, 72, 70
33	Arsenic . .	74·96	1	75
34	Selenium .	79·2	6	80, 78, 76, 82, 77, 74
35	Bromine .	79·92	2	79, 81
36	Krypton .	82·9	6	84, 86, 82, 83, 80, 78
37	Rubidium .	85·44	2	85, 87

TABLE V—*continued.*

Atomic Number.	Element.	Atomic Weight.	No. of Isotopes.	Mass-numbers of Isotopes, in orderof relative amounts in mixture.
38	Strontium .	87·6	2	88, 86
39	Yttrium .	89·0	1	89
40	Zirconium .	91	3 (4)	90, 94, 92, (96)
47	Silver . .	107·88	2	107, 109
48	Cadmium .	112·41	6	114, 112, 110, 113, 111, 116
49	Indium . .	114·8	1	115
50	Tin . .	118·7	7 (8)	120, 118, 116, 124, 119, 117, 122, (121)
51	Antimony .	121·8	2	121, 123
52	Tellurium .	127·5	3	128, 130, 126
53	Iodine . .	126·93	1	127
54	Xenon . .	130·2	9	129, 132, 131, 134, 136, 128, 130, 126, 124
55	Cæsium . .	132·8	1	133
56	Barium . .	137·4	1 (2)	138, (136)
57	Lanthanum .	138·90	1	139
58	Cerium . .	140·2	2	140, 142
59	Praseodymium	140·9	1	141
60	Neodymium .	144·3	3 (4)	142, 144, 146, (145)
80	Mercury .	200·6	6	202, 200, 199, 198 201, 204
83	Bismuth .	209·0	1	209

atomic species. Table V is supplemented by Table VI, referring to the ten radioactive elements which, together with two gaps, occupy the last twelve places in the periodic system. Of these elements, 44 species are known. Thus, in the present state of research a total of 66 out of the 89 known elements, or more than three-quarters (including all the elements with ordinal numbers from 1 to 40 and from 47 to 60), have been investigated for isotopes, and in the process 164–170 atomic species have been discovered. Since there are still 23 elements to be investigated, the total number of atomic species may well exceed two hundred.

Of the elements investigated for isotopes, 24 have been

TABLE VI

Radioactive Isotopes (including the Inactive Species of Elements Nos. 81–92)

Atomic Number.	Element.	Number of Isotopes.
81	Thallium . .	4
82	Lead . . .	8
83	Bismuth . .	5
84	Polonium . .	7
86	Radium emanation .	3
88	Radium . .	4
89	Actinium . .	2
90	Thorium . .	6
91	Protoactinium .	3
92	Uranium . .	2

proved to be single, or so-called *pure elements*, at least in the sense that, if they do contain any admixture of isotopes, the quantity of these is too small to be detected.[6] As might be expected, the atomic weights of the pure elements are very approximately integral. The small deviations can be explained on the basis of the so-called *packing effect*, which we shall discuss in a later section (§ 29).[7]

The number of isotopes is especially large in the case of xenon (9) lead (8), and tin (7 or 8). The difference in weight between the lightest and the heaviest isotope amounts in selenium, krypton, and tin, to eight units, and in xenon even to twelve units. The difference is relatively largest for lithium and neon, where the heavier isotope is respectively 16 and 10% heavier than the lighter.[8]

The instances of so-called *isobares* are also noteworthy, *viz.*, those in which atomic species differing in chemical

[6] Hydrogen and oxygen have also been proved to be pure elements by diffusion experiments.

[7] The large deviation with cæsium is, however, striking.

[8] Since the relative difference in the atomic weights of two isotopes is generally not very large, a gradual displacement of the mixture ratio in the course of time would be scarcely noticeable. For example, as a simple calculation shows, an alteration of the mixture ratio by 1% in bromine, which contains two isotopes, 79 and 81, in approximately equal quantities, would only cause a change in the mixture weight of two units in the second decimal place (0·81 − 0·79 = 0·02).

character have the same atomic weight. Such isobares occur, *e.g.*, for an atomic weight of 40 in argon and calcium, for an atomic weight of 70 in zinc and germanium, *etc.*

Since isotopes are identical in all properties determined by the electrons surrounding the nucleus, it is plain that a *separation of isotopes* can only be effected by such methods as are based on differences in nuclear properties. On the other hand, the ordinary chemical methods of separation will be of no avail, especially those based on differences in solubility and vapour pressure, in which the principal part is taken by the outer electrons. One of the main possible ways in which isotopes may be separated utilizes the fact that the *velocity of evaporation* depends on the atomic mass. *Brönsted* and *Hevesy* have actually succeeded, by this method of successive evaporations, in preparing *two species of mercury* which differ in density by 1 in 2,000, so that their mixture weights differ by 0·1 units. The same investigators, using the same process, have been able to prepare *two species of chlorine* differing in their mixture weights by more than two units in the second decimal place.[9] It is, however, too much to hope at present for a preparation of pure isotopes in this manner, for the yield becomes smaller and smaller as the separation proceeds.[10]

The differences in nuclear mass and structure between isotopes also find expression, to a small degree, in *spectral* properties. The conception of the nucleus as a point charge is only approximately correct and the electrical field surrounding the nucleus may exhibit small differences, with varying composition of the nucleus, despite equality of the nuclear charge number. To this is most probably due the fact established by *Aronberg* and by *Merton*,[11] that the wavelength of the spectrum line of *lead* at 4,058 Å.U. proves to be greater by 0·01 Å.U. for uranium lead, and smaller by 0·002 Å.U. for thorium lead, than in the case of common lead. The phenomenon of isotopy finds far stronger expression in molecular spectra than in optical spectra, as we shall later discuss at length.

[9] For further details, *vide* Hevesy and Paneth's "*Lehrbuch der Radioaktivität*" (Leipzig: Barth, 1923), Ch. xix. (English translation: Oxford, Clarendon Press).

[10] At the present time there does not yet seem to be any possibility of adapting the analytical process of the positive ray method to a method of preparation in which the ions of the isotopes generated in the discharge tube might be separated, in virtue of their different deflections in both an electric and a magnetic field, and then collected.

[11] T. R. Merton, *Proc. Roy. Soc.*, *A* **100** (1921), p. 84.

§ 24. Disintegration of the Nucleus, and Transformation of the Elements.

Some few years after the first observation of radioactive radiation, closer investigation led to a new discovery of far greater moment, *viz.*, the discovery of the *transformation of elements.* In 1900 *Rutherford* established the surprising fact that the radioactive element thorium is constantly evolving a gas that is itself radioactive, and which he called *thorium emanation* [1]; and a similar process was shortly afterwards discovered by *Dorn* to be going on with *radium.* We know now that thorium emanation and radium emanation are isotopes of the rare gas whose nuclear charge number is 86; the elementary nature of the emanation was manifest from its *characteristic spectrum.*

In 1902 an explanation of the formation of the emanations was put forward by *Rutherford* and *Soddy* in a theory distinguished by great simplicity, but representing a complete break with the hitherto deeply rooted and fundamental conceptions of physics and chemistry. Rutherford and Soddy's theory, which is known as the *disintegration theory,* is in fact based on the assumption that the atoms of radioactive substances undergo a series of successive disintegrations; and that the expulsion of the projected particles in the form of radioactive corpuscular radiation changes the chemical character of the residual atom.[2]

From the point of view of the nuclear theory of the atom later developed, we must obviously refer the processes of disintegration to the *atomic nucleus* itself; and then, since the *nuclear charge number* determines the *chemical character,* the phenomenon of the *transformation of radioactive elements*

[1] Rutherford's discovery resulted from the fact, previously observed by M. and Mme. Curie, that bodies enclosed in the same space as radium themselves become radioactive. This phenomenon, which at the time was termed "induced radioactivity," was more fully investigated by Rutherford in the case of thorium, and it was then that he discovered that the "induced activity" is due to the emanation.

[2] The disintegration theory thus really opened up absolutely new territory in natural philosophy, for, in the processes of nature previously allocated to physics, the molecules remain unchanged, whilst chemical processes were defined as being those which occur inside the molecules, without, however, producing any changes in the atoms. But the phenomena of radioactivity were referred by Rutherford's disintegration theory to processes occurring *inside the atoms themselves,* and thereby altering them. Moreover, radioactivity was shown to be a property of the atoms themselves by the fact that it was found to be quite independent of the kind of chemical compound concerned, and also that it is entirely unaffected by temperature changes of a thousand degrees, whereas the velocity of most chemical reactions is doubled by a rise in temperature of only ten degrees.

at once becomes intelligible. The expulsion of an α- or a β-particle by the disintegrating atom distinguishes between two varieties of transformation, known briefly as the α- and the β-transformation. Since an α-particle possesses a positive charge of two elementary quanta and a mass of four hydrogen atoms (§ 4), an *alpha-transformation reduces the atomic number by two*, and the *atomic weight by four*. On the other hand, since β-particles are identical with electrons, a *beta-transformation raises the atomic number by one, without noticeably altering the atomic weight*. Hence, in the *periodic system of the elements*, an α-transformation *displaces* the element affected *through two places to the left*, whilst a β-transformation displaces it *through one place to the right.*

Actually, the emanation resulting, together with α-rays, directly from radium is found to have an atomic weight four units lower than that of radium,[3] *viz.*, 222 as compared with 226 ; and, whereas radium belongs to the second group of the periodic system, *i.e.*, the alkaline earth group, the emanation (a rare gas) belongs to the zero group, or, what amounts to the same thing, to the eighth group.

The *formation of helium during the disintegration of the emanation, which is itself radioactive*, was first demonstrated in 1903 by *Ramsay* and *Soddy*, using a *spectroscopic method*. This was the first instance of the observation of the generation from another element of an element already known to chemists. The actual *identity of the atoms of helium formed* and the *α-particles* expelled during the disintegration of radioactive substances (at least, after neutralization) was proved in 1909 by a beautiful experiment carried out by *Rutherford* and *Royds*.

In this experiment, a substance emitting α-rays was contained in a closed glass tube, the wall of which was not thicker than about one-hundredth of a millimetre, so that the α-rays could pass through it. This tube was placed inside another, thick-walled, glass tube, which was likewise closed. Since no α-rays could be detected outside the larger tube, it was evident that all the α-particles passing through the thin glass wall were retained in the evacuated space between the two walls. In this evacuated space, however, Rutherford and Royds were actually able to demonstrate spectroscopically the presence of helium, after some time had elapsed.

On closer investigation, it was found that a preparation

[3] The density, and hence the atomic weight, of radium emanation were first determined by Gray and Ramsay in 1902, a measured volume of it being weighed by means of a special micro-balance.

containing one gram of radium (*e.g.*, in the form of 1·314 grams of $RaCl_2$) generates in a year about 167 mm.³ of helium, corresponding to a mass of 0·0000298 gram (the density of helium being taken as 0·000179).[4] Since the mass of a helium atom is equal to the atomic weight (4·00) divided by the Loschmidt number, we obtain the number of helium atoms generated in a year by one gram of radium, on dividing the above-given number (0·0000298) by four, and then multiplying by the Loschmidt number ($6{\cdot}062 \times 10^{23}$).[5] In this way we find that one gram of radium produces $4{\cdot}52 \times 10^{18}$ helium atoms in a year.

Again, the accurate counts already mentioned (in § 4), *made by the electrical method*, have shown that a gram of radium emits $3{\cdot}72 \times 10^{10}$ α-particles in one second.[6] This number is four times as large if the preparation is in equilibrium with three disintegration products which are formed from radium, and also emit α-rays [*cf.* the remarks in § 25 in connection with eqn. (12)]. Since there are 86,400 seconds in a day, and 365 days in a year, the number of α-particles ejected by the radium preparation in a year is equal to $3{\cdot}72 \times 10^{10} \times 4 \times 365 \times 86{,}400$, or $4{\cdot}69 \times 10^{18}$. The good agreement between the calculated number of the α-particles emitted and the accumulated helium atoms formed an important empirical confirmation of the disintegration theory. Conversely, the observation of the *rate of formation of helium* from radium, together with a *count of the α-particles*, enables the *Loschmidt number* to be *determined directly*, although not very accurately.

In every radioactive substance, the ratio of the number of atoms disintegrating in a second to the total number of atoms of the substance present is found to represent a characteristic constant. This ratio is called the *disintegration constant*, on the essential assumption that the *number of atoms disintegrating* in an element of time is always *proportional to the number of atoms available*. Denoting the disintegration constant by λ and the number of atoms present by N, we therefore have for the number of atoms disintegrating in an element of time (Δt) the relation

$$\Delta N = -\lambda N \Delta t. \qquad (1)$$

[4] The density of monatomic helium is only twice that of diatomic hydrogen.

[5] Since, according to the well-known Avogadro's Law, the number of molecules in 1 cm.³ at normal pressure and 0° is the same for all gases, *viz.*, $2{\cdot}708 \times 10^{19}$, the number of monatomic helium molecules contained in 0·167 cm.³ is found by simply multiplying together the two numbers just given.

[6] The exact determination of this number is due to Hess and Lawson (1918).

On replacing the differences by differentials and integrating, we obtain the equation

$$N = N_0 e^{-\lambda t}, \tag{2}$$

where N_0 denotes the number of atoms at the time $t = 0$. The time (T) in which N diminishes to one-half of N_0, half of the atoms originally present having by then disintegrated, is known as the *half-value period*, or period of half change. Its connection with the disintegration constant follows from the relation

$$\frac{1}{2} = e^{-\lambda t};$$

whence

$$T = \frac{\log_e 2}{\lambda} = \frac{0{\cdot}693}{\lambda}. \tag{3}$$

The reciprocal value of the disintegration constant is called the *period of average life;* it is larger than the half-value period, being obtained on dividing the latter by 0·693.

The disintegration constant of *radium* may at once be calculated from the number of α-particles emitted in a second. For we obtain the number of atoms in one gram of radium simply by dividing the Loschmidt number by the atomic weight of radium (226). This gives $2{\cdot}68 \times 10^{21}$; and, since $3{\cdot}72 \times 10^{10}$ α-particles are emitted in one second, while the same number of radium atoms disintegrate, the value resulting for the disintegration constant of radium is

$$\lambda\,(\mathrm{Ra}) = \frac{3{\cdot}72 \times 10^{10}}{2{\cdot}68 \times 10^{21}} = 1{\cdot}39 \times 10^{-11}\ \mathrm{sec.}^{-1}. \tag{4}$$

It follows from eqn. (3) that the *half-value period of radium* is given by

$$T\,(\mathrm{Ra}) = 4{\cdot}99 \times 10^{10}\ \mathrm{sec.} = 1{,}580\ \mathrm{years}. \tag{5}$$

For many radioactive substances, the half-value period (to values as low as about 0·002 sec.) can also be derived from observations on the *diminution in intensity of the radioactive radiation effects.* Thus, denoting the intensity of the radioactive radiation measured by any apparatus[7] at an arbitrary time and at the time $t = 0$ by I and I_0 respectively, we have, as is also found experimentally, the relation

$$I = I_0 e^{-\lambda t}. \tag{6}$$

In the case of radium emanation, the value found in this

[7] The simplest method is to measure the conductivity acquired by air as a consequence of the ionizing effect of the radioactive radiation.

way for the exponential constant is precisely the same as that given for the disintegration constant by the scintillation method. At the same time, this evidently proves that the intensity of the radioactive radiation at any moment is proportional to the number of atoms of the radioactive substance in existence at that moment.

§ 25. Radioactive Equilibrium.

In the following considerations, we shall apply the fundamental equation of the disintegration theory [§ 24 eqn. (2)] to the special case in which the *product of the transformation* is itself radioactive, but with a much *smaller life period than the parent substance.* Such, for example, is the case with radium emanation and radium, the half-value periods being in the approximate ratio 1 : 150,000.

Let λ be the disintegration constant of the parent substance, and λ' that of the transformation product. We shall suppose N atoms of the former and N' atoms of the latter to be present at a given moment. At an arbitrary time $t = 0$, let the values of N and N' be represented by the constants N_0 and N_0'.

We now consider an arbitrarily chosen element of time Δt; during that interval, atoms of the transformation product will *disintegrate* and, *simultaneously,* new atoms of the product will *be generated* by the disintegration of the parent substance. The number of disintegrating atoms is $\lambda' N' \Delta t$; the number of newly generated atoms $\lambda N \Delta t$.

If now, as we assumed, the parent substance is very long-lived as compared with the transformation product, then, even after an interval considerably longer than the half-value period of the transformation product, the quotient N/N_0 will not differ noticeably from unity.[1] Hence, when we are considering the state of the transformation product, we may always replace the variable quantity N by the constant N_0. The change in the number of the atoms of the transformation product in an element of time (dt) is accordingly given by the relation

$$(1) \qquad \frac{dN'}{dt} = -\lambda' N' + \lambda N_0 .$$

In order to integrate this equation, it is best to begin by differentiating again with respect to the time; denoting the

[1] In a system composed of radium and emanation we might take one year as the time.

differential coefficient dN'/dt by φ, for the sake of brevity, we then obtain

$$\frac{d\varphi}{dt} = -\lambda'\varphi. \tag{2}$$

It immediately follows from this that (replacing φ by the derivative dN'/dt)

$$\frac{dN'}{dt} = C \cdot e^{-\lambda' t}. \tag{3}$$

By means of eqn. (1), we can easily determine the value of the constant C, for, at the time $t = 0$, it must be equal to the differential coefficient dN'/dt. In accordance with eqn. (1) we have

$$\left|\frac{dN'}{dt}\right|_{t=0} = -\lambda' N_0' + \lambda N_0. \tag{4}$$

The right-hand side of this equation therefore represents the constant C, so that eqn. (3) may be written in the form

$$\frac{dN'}{dt} = (-\lambda' N_0' + \lambda N_0) e^{-\lambda' t}. \tag{5}$$

Integrating this again, we have

$$N' = N_0' e^{-\lambda' t} - \frac{\lambda}{\lambda'} N_0 e^{-\lambda' t} + C'. \tag{6}$$

The integration constant C' can be determined from the equality of the numbers N' and N_0' at the time $t = 0$. Introducing the time $t = 0$ into eqn. (6), we find

$$N_0' = N_0' - \frac{\lambda}{\lambda'} N_0 + C', \tag{7}$$

whence

$$C' = \frac{\lambda}{\lambda'} N_0. \tag{8}$$

Eqn. (6) may therefore be written in the form

$$N' = N_0' e^{-\lambda' t} + \frac{\lambda}{\lambda'} N_0 (1 - e^{-\lambda' t}). \tag{9}$$

In this equation let us give t a value which is large compared with the half-value period of the transformation product, but small compared with the half-value period of the parent substance. For such a value of t, the expression $e^{-\lambda' t}$ may simply be put equal to zero. Eqn. (9) then becomes

$$N' \lambda' = N_0 \lambda. \tag{10}$$

If, however, this relation be satisfied, then by eqn. (1)

$$\frac{dN'}{dt} = 0. \tag{11}$$

The number of atoms of the transformation product must therefore be constant because, in a given time, just as many new atoms of the transformation product are generated as disintegrate. The system consisting of the parent substance and its transformation product is then said to be in *radioactive equilibrium.* Even if equilibrium did not exist at the time $t = 0$, it must nevertheless be attained after an interval which is large as compared with the half-value period of the transformation product, provided that, during this interval, there be no external interference with the system (such as the removal of the transformation product as formed).

We shall now suppose the transformation product in its turn to generate a new active substance, the number of atoms and the disintegration constant of which may be denoted by N'' and λ'', this substance again generating another active product, and so on. Then, when radioactive equilibrium exists, the following relation must hold between the parent substance and all the transformation products :—

$$N_0 \lambda = N' \lambda' = N'' \lambda'' = N''' \lambda''', \textit{etc.} \tag{12}$$

Since the number of atoms of a radioactive substance disintegrating in unit time is given by the product of the disintegration constant and the number of atoms present, *equal numbers of atoms of all substances* which are in *radioactive equilibrium* with each other *disintegrate* in the same time. This explains why a radium preparation ejects *four times* as many α-particles as radium itself, for it contains, besides radium, three transformation products emitting α-rays (RaEm, RaA, RaC), in equilibrium with the radium.

Eqn. (12), moreover, permits us to derive the *disintegration constant of the long-lived parent substance* from the known disintegration constant of any transformation product. For example, it has been found that, in all uranium minerals, there is a content of $3{\cdot}4 \times 10^{-7}$ grams of *radium* for every gram of *uranium*. Since the atomic weights of uranium and radium are 238 and 226 respectively,[2] it follows that, in radioactive equilibrium, corresponding to n-atoms of uranium, we have $n \times 3{\cdot}4 \times 10^{-7} \times 238/226$, or $n \times 3{\cdot}6 \times 10^{-7}$ atoms of radium. We then obtain the half-value period of uranium,

[2] It is not necessary to take into consideration uranium II, which is always contained as an isotope in the mixed element uranium, for it only forms about 0·025 % of the whole element.

on dividing that of radium (1,580 years) by $3{\cdot}6 \times 10^{-7}$; the *half-value period of uranium* is thus found to be 4,400 million years. This value agrees well with that obtained directly by the scintillation method. Counts have shown that one gram of uranium emits $2{\cdot}3 \times 10^{4}$ α-particles in a second, only half of which, however, arise from uranium I, the other half being generated by uranium II, which, as we shall see in the next section, is always present in radioactive equilibrium with uranium I, though in much smaller amount. Since the atomic weight of uranium is 238, one gram contains about $2{\cdot}55 \times 10^{21}$ atoms. The disintegration constant is therefore found on dividing the half of $2{\cdot}3 \times 10^{4}$ by this number, the result being

$$\lambda\,(\mathrm{U}) = 4{\cdot}5 \times 10^{-18}\ \mathrm{sec.}^{-1}.$$

It follows from this, according to § 24 eqn. (3), that the half-value period of uranium amounts to $4{\cdot}9 \times 10^{9}$ years, in good agreement with the value derived indirectly.

The ratio between the half-value periods of *radium* and *radium emanation* is also found to obey eqn. (12) satisfactorily. One gram of radium is in equilibrium with $0{\cdot}663$ mm.^{3}, or $6{\cdot}56 \times 10^{-6}$ grams of emanation, this quantity, which serves as a unit of measure, being known as a "*curie*."[3] The half-value period of emanation must hence be equal to that of radium multiplied by $6{\cdot}56 \times 10^{-6}$, and by the ratio between the atomic weights of radium and the emanation, *i.e.*, by 226/222.[4] In accordance with § 24 eqn. (5), we thus find for the half-value period of radium emanation $3{\cdot}33 \times 10^{5}$ sec., while direct observation on the decrease in the intensity of the radiation gives a value of $3{\cdot}29 \times 10^{5}$ sec., or 3·81 days.

Finally, we will apply eqn. (9) to the special case in which, at a given moment which we choose as zero on the time scale, the *transformation product is separated from the longer-lived parent substance* with which it is supposed, till then, to have been in equilibrium. From the moment of separation, we have therefore to distinguish between two preparations. In the first, the transformation product which was completely removed at the time $t = 0$ continues to form, whilst in the

[3] The unit of measure used for radium emanation concentrations is the *Mache* unit; this corresponds to an emanation content of $3{\cdot}64 \times 10^{-10}$ curie/litre. (The quantity of emanation then contained in one litre will produce, on transference to an electroscope, a saturation current of 10^{-3} electrostatic units per second.) The strongest therapeutic springs have an emanation content of 2,000 or 3,000 Mache units.

[4] For n radium atoms are in equilibrium with $n \times 6{\cdot}56 \times 10^{-6} \times 226/222$ emanation atoms.

second preparation, which contains no parent substance, the number of atoms of the transformation product in it decreases continuously, in consequence of their progressive disintegration. Let N_1' and N_2' be the numbers of atoms of the disintegration product in the two preparations at any given time.

It follows from eqn. (9) that, since N_1' is to be made equal to zero at the time $t = 0$, the number N_1' is given by

$$(13) \qquad N_1' = \frac{\lambda}{\lambda'} N_0 (1 - e^{-\lambda' t}).$$

On the other hand, at the time $t = 0$ the number N_2' is equal

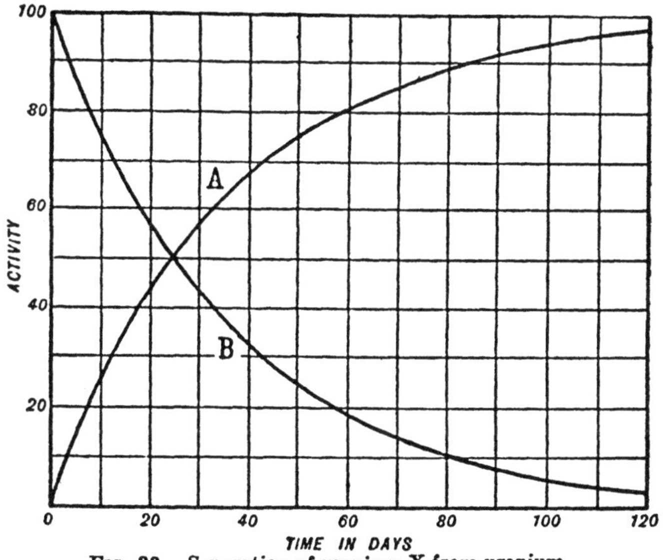

FIG. 33.—Separation of uranium X from uranium.

to $N_0 \lambda/\lambda'$, because of the equilibrium existing up to that time. Hence, at an arbitrary time [in accordance with § 24 eqn. (2)],

$$(14) \qquad N_2' = \frac{\lambda}{\lambda'} N_0 e^{-\lambda' t}.$$

Adding eqns. (13) and (14), we find

$$(15) \qquad N_1' + N_2' = \frac{\lambda}{\lambda'} N_0;$$

i.e., the *sum of the atoms of the transformation product in the two preparations* is *not affected by the separation*, but remains *constant*.

An instructive example of the fulfilment of this law is afforded by the *separation of uranium X from uranium.* Uranium X is of far shorter life than uranium, its half-value period only amounting to 23·8 days.[5] Since uranium emits α-rays, but uranium X β-rays, the intensity of the β-radiation emitted by the separated preparations forms a measure of the number of uranium X atoms present.[6] In Fig. 33 curve A shows the *increasing* intensity of the β-radiation from the "recovering" uranium preparation, and curve B the *decreasing* intensity in the separated uranium X preparation. At every instant the sum of the ordinates of the two curves has the same value, namely, the initial value (supposed equal to one hundred).

§ 26. The Transformation Series.

At the present time, 40 radioactive elements,[1] or, as they are briefly termed, "*radioelements*," with the most various life periods are known in physics; the half-value periods range from 10^{-11} sec. to 10^{12} years. Apart from potassium and rubidium, from which a weak β-radiation has been detected, radioelements are distributed over the last twelve places in the periodic system with the ordinal numbers 81 to 92, omitting, however, the 85th and 87th places.

Exact investigations have shown that the radioelements (apart from potassium and rubidium) can be arranged in *three transformation series*, in such a way that each element in such a series arises from the preceding element through either an α- or a β-transformation. The three series are known as the uranium-radium series, the actinium series, and the thorium series, though the second is obviously only a branch of the first.

The *uranium-radium series* is represented in Table VII, in which elements with the same atomic weight are situated in the same row, and elements with the same atomic number in the same column; the groups of the periodic system corresponding to the atomic numbers are also shown. An α-transformation, in which the atomic weight decreases by four, is indicated by an oblique arrow, and a β-transformation, in which the atomic weight remains unaltered, is shown by a horizontal arrow. The half-value period is given under each element (here, and in the two following tables, years,

[5] The small admixtures of uranium X_2 and uranium II can be disregarded, and we may speak simply of uranium and uranium X. *Cf.* § 26.

[6] The α-radiation can be cut out by absorption.

[1] Not including the stable end products of the transformations.

Table VII
The Uranium-Radium Series

Atomic weights	81 III B	82 IV B	83 V B	84 VI B	85 VII B	86 VIII B	87 I A	88 II A	89 III A	90 IV A	91 V A	92 VI A
238												UI $4{\cdot}4 \times 10^{9}$ y.
234										UX_1 23·8 d.	UX_2 1·14 m.	UII 1×10^{8} y.
230										Io 9×10^{4} y.		
226								Ra 1580 y.				
222						RaEm 3·81 d.						
218				RaA 3·05 m.								
214		RaB 26·8 m.	RaC 19·5 m.	RaC′ $1{\cdot}5 \times 10^{-8}$ s.								
210	RaC″ 1·32 m.	RaD 16 y.	RaE 4·98 d.	RaF 136·5 d.								
206		RaG stable										

days, hours, minutes, and seconds are denoted respectively by the letters *y*, *d*, *h*, *m*, and *s*).

As can be seen from Table VII, *uranium* is a *mixed element* with *two isotopes*, the heavier of which undergoes one α- and two β-transformations to form the lighter. One gram of uranium I, however, is in equilibrium with only 0·25 mgm. of uranium II, and, indeed, with only $1{\cdot}5 \times 10^{-11}$ grams of uranium X_1, and $4{\cdot}9 \times 10^{-16}$ grams of uranium X_2. Hence from the point of view of practical chemistry, the mixed element can be regarded as being pure uranium I. Uranium II cannot be separated from uranium I, with which it is isotopic; on the other hand, uranium X_1 and uranium X_2, which are isotopes of thorium and protoactinium, can be separated by suitable precipitation methods.

An α-transformation leads from uranium II to the long-lived parent substance of radium, *ionium*. Inasmuch as this is an isotope of thorium, it cannot be prepared in a pure state, but only in the form of an enriched thorium preparation, to the extent of 30%.[2] Radium, which has already been obtained in some hundreds of grams from minerals, and which has also been prepared in a chemically pure state, gives rise to *radium emanation* on an α-transformation. Since this is a rare gas, not chemically combined but only occluded in radium salts, it can easily be liberated on heating.

Radium emanation produces the so-called *short-lived active deposit of radium*, consisting of five substances, RaA, RaB, RaC, RaC′ and RaC″, whose genetic relations can be seen from Table VII. These substances are all solid, and are mostly charged positively immediately on their production, so that they are readily deposited on negatively charged plates. The five elements are so short-lived that they can only be observed in mixtures, though the proportions of the mixtures may be varied at will. Analyses of the curves representing the decrease in activity of such mixtures have made it possible to determine the half-value periods of the individual constituents of the short-lived active deposit. The smallest of the half-value periods have been indirectly determined on the basis of the relation between the life period and the range of the emitted α-rays, which we shall discuss in the following section.

At radium C, the transformation series *branches*, a β-transformation leading to the extraordinarily short-lived RaC′, whereas an α-transformation leads to RaC″. The α-trans-

[2] Whereas the atomic weight of pure thorium is 232·1, the combining weight of the thorium preparation richest in ionium is 231·5.

formation, however, is much rarer than the β-transformation; of 10,000 RaC atoms, only 4 undergo the transformation into RaC″.

When a sufficiently large quantity of emanation (such as half a cubic millimetre) disintegrates inside a closed glass vessel, a fine deposit which has the properties of a *species of lead* (known as *radium D*) is formed on the walls. The transformation law shows that radium D must result from an α-transformation of RaC′ (which emits α-rays), and also by a β-transformation from RaC″. Radium D has the comparatively large half-value period of 16 years; it changes after emitting a weak β-radiation into an isotope of bismuth, known as radium E, from which another β-transformation gives rise to RaF, known as *polonium*. Polonium emits α-rays, and when it has decayed no new rays can be detected. Hence the element produced from polonium after an α-transformation and known as RaG, which (according to the transformation law) must be a *species of lead*, representing the so-called *uranium lead*, must be assumed to be the *end product of the uranium-radium series.*

Quite analogous to the radium series is the *actinium series*, represented in Table VIII. There is no doubt about the sequence from protoactinium to the presumably final product actinium D, isotopic with lead. *Protoactinium* is a very long-lived element, with a half-value period of 12,000 years; the disintegration product, arising from an α-transformation, is *actinium*, whose half-value period is much smaller, amounting only to 20 years. Hitherto it has not been possible to prepare either protoactinium or actinium in a pure state, for they have not yet been completely separated from their homologues, tantalum and lanthanum. One β- and two α-transformations lead from actinium to *actinium emanation*, whose life period, however, only amounts to a few seconds. Actinium emanation gives rise to the *active deposit* of actinium. The sequence of transformations in this deposit much resembles that of the radium active deposit. Whereas, however, almost the whole of the RaC atoms undergo a β-transformation, just the converse is the case with the atoms of AcC. Of 10,000 AcC atoms, about 9,968 emit α-rays and form AcC″, whilst only about 32 undergo the β-transformation into AcC′. AcC′ and AcC″ lead directly to the final stable product, AcD, whereas the corresponding product radium D is weakly radioactive.

Evidence that the *actinium series* probably represents a *branch of the uranium series* is to be found in the observed

TABLE VIII
The Actinium Series

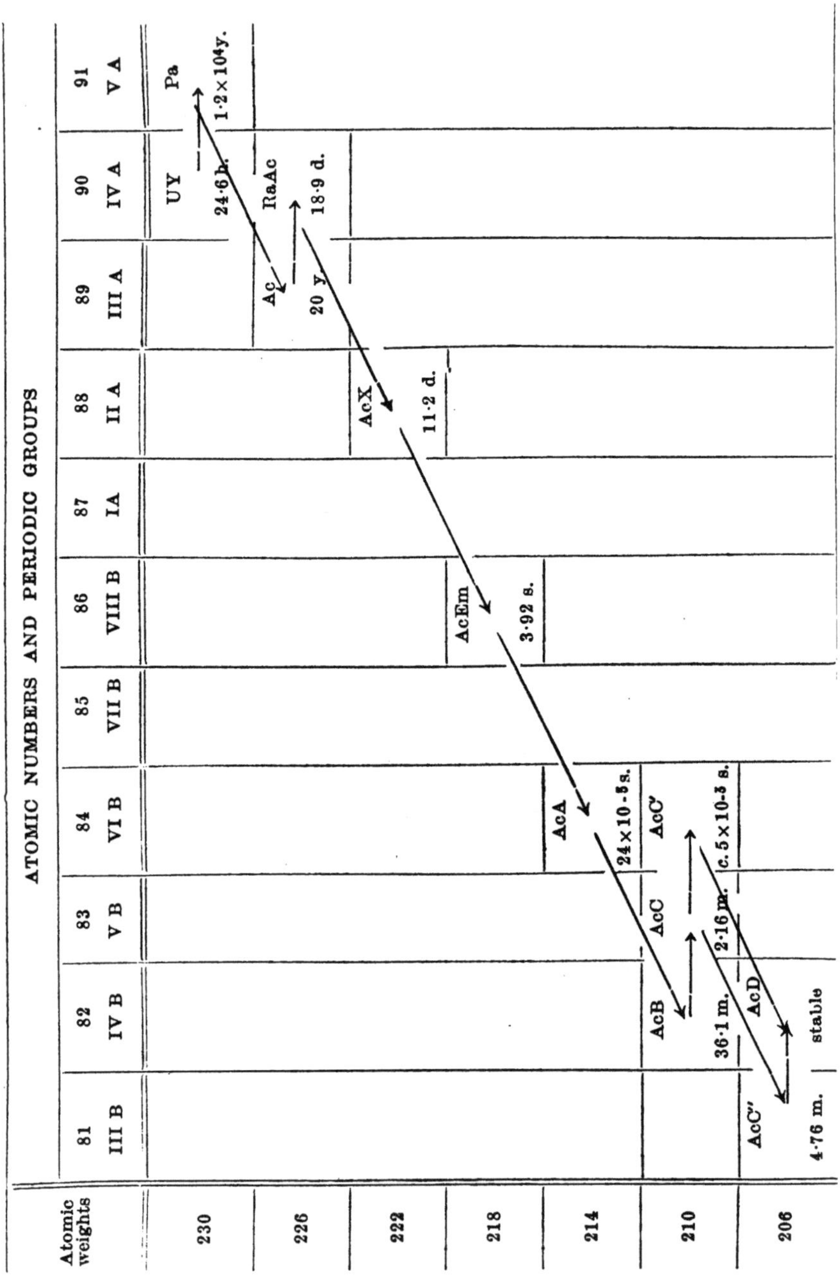

fact that the analogous members of the actinium series occur in all uranium minerals to the same extent, being responsible for about three *per cent.* of the total activity. This means that the uranium series branches in such a way that 97 out of every 100 atoms follow the transformation into radium, and 3 the transformation into actinium. At what point the branching occurs is, however, not yet certain. On the grounds of the transformation law alone, it is probable that the parent substance of protoactinium is *uranium Y*, an isotope of thorium, emitting β-rays, which is present in uranium and contributes three *per cent.* towards the activity, though the relative masses are only in the ratio of 2×10^{-14} to 1. It is further possible that uranium Y, like ionium, is produced from uranium II by an α-transformation, so that a branching involving two α-transformations would occur at this radio-element. The atomic weights of the actinium series are also only hypothetical, for no single member of the series has yet been isolated in the pure form in sufficient quantity to allow of an experimental determination of its atomic weight.[3]

Furthermore, a branching in the uranium series gives rise to *uranium Z.* Of 10,000 atoms of uranium X_1, 35 emit β-rays and form uranium Z, and this undergoes a second β-transformation into uranium II. It is an isotope of protoactinium, has a half-value period of 6·7 hours, and contributes about 0·35 per cent. towards the activity of uranium, in which it is always present, though the mass ratio is only 5×10^{-16} to 1.

The complete *independence of the thorium series from the uranium series* is indicated by the fact that both uranium minerals almost free from thorium, and thorium minerals almost free from uranium, are known. The thorium series is shown in Table IX.

From the initial substance of the thorium series, *thorium*, an α-transformation leads to the relatively long-lived *mesothorium* 1. Since its half-value period amounts to 6·7 years, and in all thorium minerals there are about 3×10^{-10} grams of mesothorium 1 to 1 gram of thorium, it follows from the law of radioactive equilibrium that the half-value period of thorium must be about 2×10^{10} years. Pure mesothorium 1 cannot be prepared from minerals, because it is an isotope of radium, and all thorium minerals always contain a little uranium, and thus radium.[4] The pure substance can only be

[3] Minute quantities of pure AcX have been prepared by the method of recoil (to be described in the following section—§ 27). An experimental atomic weight determination is first to be hoped for with protoactinium, as soon as a larger quantity of this has been obtained in a pure state.

[4] Since 1 gram of thorium is in equilibrium with 3×10^{-10} grams of

Table IX
The Thorium Series

Atomic weights	ATOMIC NUMBERS AND PERIODIC GROUPS									
	81 III B	82 IV B	83 V B	84 VI B	85 VII B	86 VIII B	87 I A	88 II A	89 III A	90 IV A
232										Th c. 2×10^{10} y.
228								MsTh 1 6·7 y.	MsTh 2 6·2 h.	RdTh 1·90 y.
224								ThX 3·64 d.		
220						ThEm 54·5 s.				
216				ThA 0·14 s.						
212		ThB 10·6 h.	ThC 60·8 m.	ThC′ c. 10^{-11} s.						
208	ThC″ 3·20 m.	ThD stable								

obtained in minute quantities by the disintegration of thorium salts which have been previously purified from uranium and its transformation products.

From mesothorium 1, two β-transformations[5] lead through the very short-lived mesothorium 2 to *radiothorium*, which is an isotope of thorium with an atomic weight smaller by four units, and with a half-value period of nearly two years. From radiothorium, two α-transformations lead to the extremely short-lived *thorium emanation*, which in its turn gives rise to the *active deposit of thorium*. Considerable analogy exists between this and the deposits of radium and actinium, the final product here being thorium D, which is isotopic with lead. Approximately 65% of thorium C undergoes a β-transformation and 35% an α-transformation. About one ten-thousandth of the atoms of ThC disintegrate with the emission of α-rays of abnormally long range (11·3 cm.), so that at least a three-fold branching of the thorium series must be assumed to occur at ThC.

The *final product* of all three transformation series is represented by different isotopes of *lead*. Consequently, both in uranium and in thorium minerals, appreciable quantities of lead must accumulate in the course of the ages ; and hence a determination of the lead content enables us to fix the *age of the minerals*, and therewith the age of the *geological formations* in which the minerals are found. Since $n \times 1{\cdot}5 \times 10^{-10}$ atoms out of n uranium atoms disintegrate in the course of one year, while the ratio of the atomic weight of lead to that of uranium is 206 : 238, 1 gram of uranium must produce annually $1{\cdot}3 \times 10^{-10}$ grams of uranium lead. We therefore obtain the approximate age (in years) of a mineral on multiplying the amount of uranium lead, corresponding to 1 gram of uranium, by the reciprocal of $1{\cdot}3 \times 10^{-10}$ or, roughly, by 8,000 million.[6] A ten *per cent.* content of uranium lead thus indicates an age of about 800 million years. Determinations of age must, however, be based only on the content of uranium lead (RaG), from which a common contamination of ordinary lead must be distinguished.

mesothorium, and 1 gram of uranium with $3{\cdot}4 \times 10^{-7}$ grams of radium, the quantities of mesothorium and radium contained in a thorium mineral will be approximately the same when uranium is present in the mineral to the extent of even only one in a thousand of the thorium content.

[5] It is true that no radiation can be detected from mesothorium 1 ; but since the atomic number of mesothorium 2 is 89, a β-particle must be ejected during the transformation into mesothorium 2.

[6] In an accurate calculation it must be remembered that the uranium has also decreased in amount in the course of time ; allowing for this, the ages found for the minerals are somewhat smaller.

Inasmuch, however, as the atomic weights of uranium lead and ordinary lead are different (206·0 and 207·2), a determination of the atomic weight of the mixture of RaG and ordinary lead at once enables us to calculate the proportions of the two substances.[7]

In this way it has been found that the *oldest mineral known* is a zircon from Mozambique, a 21% content of uranium lead showing the age to be 1,500 *million years.* The youngest uranium mineral appears to be a uraninite occurring in the carboniferous system, with a uranium lead content of about 4% and an age of 320 million years.[8]

Whether the uranium and thorium series, both of which presumably end with lead, begin with elements of still higher atomic weight than uranium and thorium is an open question. But it is unlikely that such higher elements can be found, for, if they were shorter-lived than uranium or thorium, then, in accordance with what has just been said about the age of minerals, they would have disintegrated long ago, seeing that a formation of new elements after the solidification of the earth is scarcely conceivable. On the other hand, if their life periods were sufficiently long, they must occur in recognizable quantities in uranium and thorium minerals, as a consequence of radioactive equilibrium, in which case they could hardly have escaped detection up till now. It does not, however, appear impossible for elements with higher atomic weights than that of uranium to exist, if they belong to transformation series other than those described, which may have hitherto eluded discovery simply because of the rarity of their members.

Members of such transformation series, as yet unknown, are possibly represented by those isotopes of *potassium* and *rubidium* to which is due the β-radiation discovered in these two metals by *Campbell* in 1907. It is true that, both with potassium and with rubidium, it is still an open question whether one of the two known isotopes is radioactive, or whether the activity is caused by a third unknown isotope. If the former be the case, it may be estimated from the hardness of the β-rays emitted that the half-value period of

[7] For example, whereas the Morogoro ore found in East Africa gives an atomic weight of 206·05, and can thus only contain uranium lead, it follows from the atomic weight of 206·4, found for the lead in uraninite from North Carolina, that 70 % of the lead has been formed from uranium, and 30 % is due to an impurity.

[8] Thorium minerals are frequently of secondary origin, so that the ages found for them cannot always be ascribed to the geological formations in which they occur. The ages calculated for thorium minerals range from 10 million to 600 million years.

rubidium (which radiates the more strongly) is about a hundred thousand million years, while the half-value period of potassium is about a billion years.[9]

It was not until after the first observations of radioactivity in 1896 that by far the greater number of the radioelements were discovered; before that date only uranium and thorium were known, apart from potassium and rubidium. Of the radioelements discovered after that time, five represent newly discovered elements in the sense that they are the principal representatives of places in the periodic system which had previously been empty. These five are *radium* and *polonium*, discovered in 1898 by *M.* and *Mme. Curie; actinium*, discovered by *Debierne* and *Giesel* in 1899; *emanation*, discovered in 1900 by *Rutherford* and *Dorn;* and finally, *protoactinium*, discovered by *Hahn* and *Meitner* and by *Soddy* in 1918.[10] The genetic relations between the radioelements became evident on the establishment (in 1913) by *Soddy* and by *Fajans* of the law of α- and β-transformations —even before the discovery of the natural sequence of the elements.

§ 27. Radioactive Radiation.

The most striking phenomenon due to the disintegration of the nucleus, which, indeed, first drew the attention of physicists to that disintegration, is the *radioactive radiation* emitted by the disintegrating atoms, part of which is *corpuscular* in nature and part of the nature of *electromagnetic waves.* The radioactive rays arise either as so-called *primary rays from the atomic nucleus*, or as so-called *secondary rays* consequent on the interaction of the primary rays and the *electron groups*, surrounding the atomic nucleus.

Foremost among the primary rays are the α- and β-particles ejected by the disintegrating atomic nucleus. The *initial velocities of the α-particles* represent *constants* characteristic of the radioelements emitting them. Expressed in cm./sec., the initial velocities lie between $1{\cdot}40 \times 10^9$ for uranium I, and $2{\cdot}06 \times 10^9$ for thorium C′; the abnormally rapid α-rays, which, as already mentioned, are emitted by about a ten-thousandth part of the disintegrating thorium C atoms, have a velocity of $2{\cdot}3 \times 10^9$.

As was mentioned in an earlier section (§ 4), the *range* of the α-rays in any gas is *proportional to the cube of the initial*

[9] For the relation between the hardness of the β-rays and the life period, *vide* § 27.

[10] Before protoactinium, its short-lived isotope uranium X_2 (formerly called "brevium") had already been discovered.

velocity ; the same relation also exists between the distance yet to be covered at any moment and the constantly decreasing velocity at that moment. The ranges in air at atmospheric pressure and 0° likewise represent constants characteristic of the radioelements which emit α-rays ; they vary from 2·53 cm. for uranium I to 8·17 cm. for thorium C′. The exceptionally rapid α-rays emitted by a ten-thousandth part of the disintegrating thorium C atoms have a range of 11·3 cm. Furthermore, the range varies directly as the absolute temperature of the gas, and inversely as its pressure, so that the range of α-rays *in vacuo* is found to be extremely large. Referred to RaC′, the range in air at 0° and atmospheric pressure is connected with the initial velocity by the relation

(1) $$v^3 = 1{\cdot}076 \times 10^{27} R_0 ,$$

or,

(2) $$v = 1{\cdot}025 \times 10^9 R_0{}^{1/3} .$$

The range at 15°, which is frequently given, is obtained from R_0 on multiplying by 1·055.

A simple relation connects the ranges of α-rayers with their *life-periods.* As was found by *Geiger* and *Nuttall* in 1911, the *logarithms of the disintegration constants* are *linear functions of the logarithms of the ranges,* within each transformation series. We have the relation

(3) $$\log \lambda = A + B \log R_0 ,$$

where B is a universal constant, but A has a definite value for each of the three transformation series (*cf.* Fig. 34). With logarithms to the base 10, B is equal to 60 and, for the uranium-radium series, A is equal to (— 41·5). By means of eqn. (3) it has been possible to determine the half-value periods of the exceedingly short-lived radioelements. That eqn. (3) is not generally valid, however, can be seen from the position of AcX in Fig. 34, the range corresponding to the thorium rather than to the actinium series.[1]

Whereas the ranges are taken as characteristic constants for the radioelements emitting α-rays, the *absorption coefficients in aluminium* are generally taken for those elements which emit β-rays. Since the intensity of the β-rays diminishes in accordance with an exponential law, as experience shows, it decreases from I_0 to I, on passing through a layer of thickness d, according to the relation

(4) $$I = I_0 e^{-\mu d} ,$$

[1] As Fig. 34 shows, Geiger and Nuttall's equation does not hold in general for the thorium and the actinium series nearly so accurately as for the uranium-radium series.

where μ denotes the absorption coefficient. The value of d for which I is equal to half of I_0 is known as the *half-value thickness*, or half-value layer; it is connected with the absorption coefficient by the relation [2]

$$D = \frac{\log_e 2}{\mu}. \tag{5}$$

Expressed in reciprocal centimetres, the absorption co-

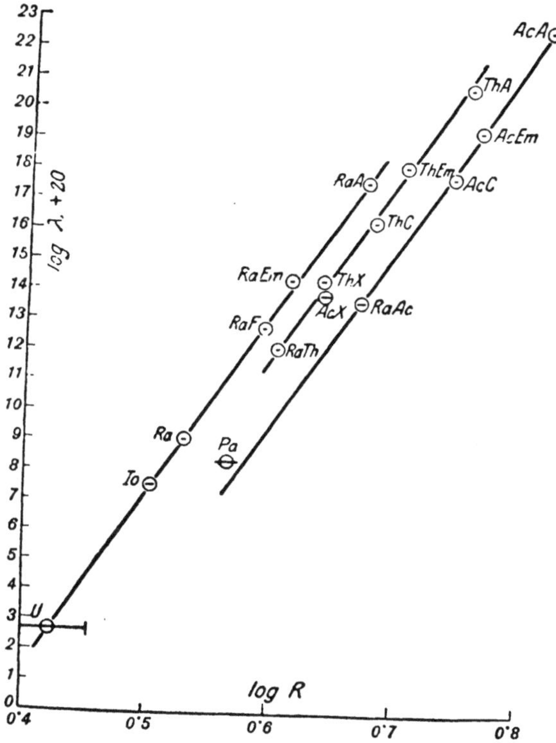

FIG. 34.—Relation between range and disintegration constant. (*Zeitschr. f. Phys.*, vol. 8. Springer, Berlin.)

efficients in aluminium lie between 13 (for RaB and RaC) and 5,500 (for RaD); the half-value thicknesses between $5{\cdot}13 \times 10^{-2}$ and $1{\cdot}26 \times 10^{-4}$ cm. Naturally, the smaller the absorption coefficient and the larger the half-value thickness, the harder and thus the more penetrating are the rays.

[2] The same relation exists between half-value thickness and absorption coefficient as between half-value period and disintegration constant.

Many radioelements possess *several different values for the absorption coefficient*,[3] which shows that the β-radiation emitted by them is composed of rays of different velocities. This is still more clearly demonstrated by the *magnetic spectra*, which are based on the different deflections of rays of different velocities in a magnetic field. Thus, if the β-radiation as deflected by a magnet be allowed to fall on a photographic plate (which is thereby blackened), with a suitable arrangement of the apparatus sharp black lines will be formed on the plate (Fig. 35). Each of these lines corresponds to a definite initial velocity, which can be easily calculated,[4] but some of them are due to the secondary β-rays (which we shall discuss later), only those of higher velocity being of primary origin.[5] The initial velocities of the β-rays lie between $9{\cdot}9 \times 10^9$ and $2{\cdot}994 \times 10^{10}$ cm./sec., amounting to between 33 and 99·8% of the velocity of light. The fastest β-rays have been found in the magnetic spectrum of radium C. In general, large values of the initial velocity are associated with small half-value periods.[6]

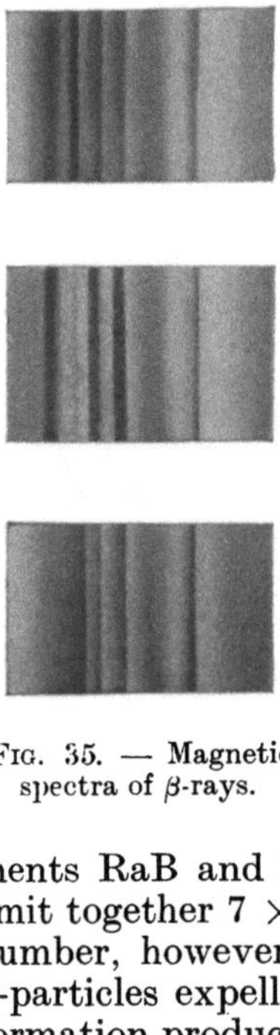

FIG. 35. — Magnetic spectra of β-rays.

By measuring the charge imparted by β-rays to an electroscope, the *number of β-particles emitted in a second* by a β-rayer can be calculated. It is found that this number is always *equal to the number of atoms undergoing a β-transformation* in a second. For example, it has been shown that the quantities of the β-raying elements RaB and RaC in equilibrium with a gram of radium emit together 7×10^{10} β-particles in a second. Half of this number, however, is nearly identical with the number of β-particles expelled by a gram of radium (apart from transformation products), which, as already mentioned, amounts to $3{\cdot}72 \times 10^{10}$. (The agreement is not perfect, because the

[3] *E.g.*, radium B has three μ-values : 890, 77, and 13·1.

[4] *Cf.* § 3 eqn. (1).

[5] The same applies to widely differing values of the absorption coefficient.

[6] Efforts to establish a quantitative relation between the disintegration constant and the absorption coefficient (on the lines of Geiger and Nuttall's equation) have so far been unsuccessful, principally because an exact separation of primary from secondary lines in the magnetic β-ray spectrum has not yet

primary β-particles are accompanied by secondary particles.) The emission of negative β-particles produces on the preparation ejecting them strong *positive charges*, up to 150,000 volts with a suitable experimental arrangement.

The number of γ-light-quanta emitted by a radioactive preparation in a given time is found to be equal to the number of atoms which disintegrate in the same time,[7] Like β-rays, γ-rays are also generally characterized by their *absorption coefficients in aluminium ;* the hardest γ-rays (from thorium C″) have an absorption coefficient of 0·0916 cm.$^{-1}$, while the softest rays (from ionium) have a coefficient of 1,088 cm.$^{-1}$. The half-value thickness in air for the hardest γ-rays is some 150 metres, whereas for the fastest β-rays it is only about 1 metre.

In the case of soft and moderately hard γ-rays, the *wave-lengths* can be determined by the crystal method (*cf.* § 19). Thus, more than 20 separate lines have been detected in the combined γ-ray spectrum of the mixture of RaB and RaC, the wave-lengths lying between 71 and 1,365 X.U. It has only been possible to determine smaller wave-lengths indirectly, values as low as 20 X.U. having been calculated ; this will be discussed more fully in a later section (§ 31). Estimations lead to a probable wave-length of only 5 X.U. for the hardest rays. It follows from Duane and Hunt's equation that the potential difference necessary to generate rays of such hardness in Röntgen tubes would be about two million volts, or nearly seven times as large as the highest potential yet produced. Only the hardest of the γ-rays, however, are likely to be primary rays.

The *secondary radioactive rays*, emanating from the portions of the atom lying outside the nucleus, are either β- or γ-rays. *Secondary β-rays* are the result of a *photo-electric* effect produced in the groups of electrons surrounding the nucleus by the primary γ-rays emitted from the latter. They are, of course, to be observed not only in such radioelements as undergo a β-transformation, but also in those which experience an α-transformation ; in point of fact, a weak β-radiation of secondary origin can be detected in the α-raying elements, radium, radioactinium, and radiothorium.[8] *Secondary γ-rays* due to primary β- or primary

[7] Kovarik [*Phys. Rev.* 23 (1924), p. 559] calculated that the amount of RaB + RaC in equilibrium with one gram of radium would emit $7{\cdot}28 \times 10^{10}$ γ-light-quanta in one second. *Cf.* R. W. Lawson [*Nature*, 116 (1925), p. 897].

[8] Secondary β-rays can probably be produced by primary corpuscular rays as well as by primary γ-rays. The so-called "δ-rays," consisting of slow electrons and emitted by all α-rayers, are very likely of this origin.

α-rays, are simply the *characteristic rays* of the elements concerned, which are also to be found in their Röntgen spectra.

A part of the corpuscular radiation emitted by radioactive substances is formed by the so-called *recoil rays*. In accordance with the Principle of the Conservation of Momentum, every atom which expels an α- or a β-particle must experience a recoil with a velocity the ratio of which to that of the expelled α- or β-particle is the same as that of the particle's mass to the mass of the remainder of the atom. For example, if an atom of radium A with an atomic weight of 218 expels an α-particle of mass 4 with a velocity of $1{\cdot}69 \times 10^9$ cm./sec., the atom of radium B resulting from the transformation must recoil with a velocity equal to that of the α-particle just given multiplied by 4/214. This velocity is about 3×10^7 cm./sec., and it is large enough to render the recoil rays recognizable.[9]

Among the effects of radioactive rays, the most striking are the *ionization of the air* and the continuous *development of heat*. The *number of pairs of ions* produced by an α-particle in the air is found to be proportional to the initial kinetic energy of the α-particle, and hence, by eqn. (2), to the square of the cube root of the range. Denoting the number of pairs of ions by k, we have (relatively to the data for radium C′)

$$(6) \qquad k = 6{\cdot}25 \times 10^4 R_0^{2/3}.$$

According to this, the long-range α-particles of radium C′ each produce some 220,000 pairs of ions; and even the slow α-particles of uranium I give rise to about 120,000. The ionization effect of β-rays is considerably less; on complete absorption of β-radiation in air, a fast β-particle produces only about 10,000 pairs of ions.[10] γ-rays also have an ionizing action. The total effect of the γ-rays emitted by any quantity of RaC is approximately as great as that of the β-rays emitted by that quantity.[11]

The ionization effect of radioactive radiation allows of the *detection of extraordinarily small and unweighable amounts* of

[9] Like α-rays, these recoil rays have a range, which for rays of RaB in air amounts to 0·14 mm. By catching the recoil rays, it has been possible to prepare some radioelements, such as AcX, in a state of purity, but only in minute quantities. The detection of recoil rays on the emission of β-particles is far more difficult, seeing that, on account of its 7,000 times smaller mass, the momentum of a β-particle, despite its 20 times larger velocity, is much smaller than that of an α-particle. Nevertheless, the recoil rays of RaC atoms produced by the β-disintegration of RaB can be observed.

[10] One β-particle in air at normal pressure generates, on an average, about 70 pairs of ions along one centimetre of its path.

[11] The ionization effect of γ-rays, referred to one centimetre of path, is in itself much less than that of β-rays. This is, however, outweighed by the much longer paths of the γ-rays.

the radioelements. The detection, and also the measurement, of such minute quantities is most simply performed by observing the time required for the leaves of a charged electroscope to collapse, when the air is made conducting by the rays from the radioactive substance.[12] By means of this method it is possible to detect even the billionth part of a gram of a radioactive substance.[13]

The *heat* which is continuously evolved by an active preparation enclosed in a vessel originates in the kinetic energy of the α- and β-rays and the recoil rays in their collision with the walls of the vessel, and also in the absorbed electromagnetic energy of the γ-rays. It may be calculated from the characteristic constants of the rays, and from the disintegration constants of the radium series, that the heat evolved in an hour by a gram of radium, together with its short-lived transformation products, must amount to 137 calories. Of this quantity of heat 90·1% is due to the α-rays, 3·4% to the β-rays, 4·7% to the γ-rays, and 1·8% to the recoil rays. In good agreement with the calculated result, the hourly development of heat is found by experiment to be 136 calories.[14]

The total quantity of heat which would be evolved by a gram of radium on complete disintegration amounts to $3{\cdot}7 \times 10^9$ calories. It is, perhaps, of interest to note that this quantity is about a million times as large as that developed on the transformation of a gram of "electrolytic gas" into water, although this chemical reaction is marked by an unusually large evolution of heat.[15]

§ 28. The Disruption of the Elements.

In radioactive phenomena the disintegration of the nucleus represents a spontaneously occurring process which

[12] Strong radioactive preparations are measured by the intensity of the "saturation currents" produced by them.

[13] *Eg.*, only about three-millionths of a milligram of radium A can be obtained from a gram of radium.—An important application of the extraordinary sensitivity of radioactive measurements has been made by *Hevesy* in his method of *radioactive indicators.* Chemical processes in which lead or bismuth take part in only very minute quantities can be detected and measured by mixing with the lead or bismuth isotopic radioelements, which cannot be separated from them, but whose presence even in the smallest amounts can clearly be recognized by radioactive measurements.

[14] The continuous evolution of heat has also been detected in the cases of polonium, thorium, and uranium. A gram of uranium in equilibrium with its disintegration products develops about a ten-thousandth of a calorie in an hour.

[15] Even if the sun were composed exclusively of uranium and its transformation products, the heat resulting from the disintegration would only account for one-half of the actual quantity of heat radiated.

cannot be influenced in any way whatsoever. It was, therefore, a discovery of the greatest importance when, in 1919, *Rutherford* for the first time demonstrated an *artificial disruption of an element,* and that an element with a low atomic number.

Rutherford's discovery was based on observations on the origin of the H-rays which we have already mentioned in an earlier section (§ 4). Shortly after the detection of these rays, *Marsden,* in conjunction with *Lantsberry,* was surprised to find that a piece of nickel foil, coated with radium C, produced *scintillations* through *ordinary air* at a distance much greater than the range in air of the α-rays emitted from the nickel foil. For, whereas the range of these α-rays (actually coming from the disintegration product RaC′) amounts only to 7 cm., scintillations could be observed even at a distance of 40 cm. This was all the more unexpected, because even H-rays generated in hydrogen have a range in air of only 29 cm.

On following up this, at first mysterious, phenomenon, Rutherford made the important discovery that the number of scintillations depended on the substance filling the space between the nickel foil and the screen. When this space was evacuated or filled with carbon dioxide or oxygen instead of with air, there were no scintillations ; on the other hand, the number increased on replacing the air by pure nitrogen. At the same time, measurements of the electric and magnetic deflections proved that the long-range rays actually were H-rays, and thus consisted of hydrogen nuclei.

These facts led Rutherford to the important conclusion that the *α-particles effect a disruption of the nitrogen nuclei, hydrogen nuclei being expelled* from the disrupted nuclei.

It is found from calculations on scintillation counts that, on an average, only one single hydrogen nucleus is liberated for some hundred thousands of α-particles emitted. Since the amount of RaC′ in radioactive equilibrium with a gram of radium ejects $3{\cdot}72 \times 10^{10}$ α-particles in a second, a preparation of one gram of radium (of which only the RaC′ is of account for the disruption of the nitrogen)[1] will thus liberate about 10^{12} hydrogen nuclei annually. Hence, the hydrogen atoms arising on neutralization would, in a year, make up a volume of about a thousandth of a cubic millimetre.

In more recent researches Rutherford and Chadwick have been able to effect the disruption also of the nuclei of the following elements (hydrogen nuclei being expelled) : F,

[1] The ranges for the other α-ray transformation products are too small.

Ne, Na, Mg, Al, Si, P, S, Cl, A, and K—thus of elements Nos. 9—19. Using other methods, Kirsch and Pettersson have produced atomic disruption in the case of Be, C, O, Ti, V, Cr, Fe, Cu, Se, and Zr.[2]

Disruption of the nuclei of elements only succeeds when the α-rays have a range of at least 5 cm. (which is the case with RaC′). That the effect of the α-particles shot into the atomic nuclei is simply one of *liberation*, however, and that the process of disruption is essentially *explosive*, is shown by the fact that the *kinetic energy of an H-particle* is found to be *larger* than that of the α-particle itself. In the case of *aluminium*, from which H-rays with a range of up to 90 cm. can be produced, the kinetic energy of the H-particles exceeds that of the α-particles by some 50%. Moreover, it is found that the H-rays travel not only in the direction of the α-rays, but in part also in the opposite direction, and this likewise supports the assumption of a nuclear explosion produced by the α-particles.

By means of the "fog-track method" mentioned in a previous section (§ 4), *Blackett*[3] has succeeded in recording photographically the process of disruption in the case of nitrogen nuclei. He photographed about 400,000 tracks of α-particles, and he actually found 8 instances in which the plainly recognizable branch track of the expelled hydrogen nucleus denoted a disruption of a nitrogen nucleus. The photographs showed, moreover, that the α-particle responsible for the disruption is captured by the nitrogen nucleus concerned, so that from the latter, through loss of a proton and gain of a helium nucleus, is produced an obviously unstable isotope of oxygen, with an atomic weight of 17.[4]

§ 29. Structure of Atomic Nuclei.

In conjunction with the radioactive β-radiation, the H-rays released by nuclear disruption afford a convincing

[2] *Cf.* Rutherford, "Disintegration of Atomic Nuclei," *Nature*, 115 (1925), pp. 493-4: Pettersson and Kirsch, "*Atomzertrümmerung*" (Leipzig, Akademische Verlagsgesellschaft, 1926).

[3] *Proc. Roy. Soc. A.* 107 (1925), p. 349.

[4] In 1925 much sensation was caused by the reports of Miethe that he had apparently effected the transmutation of mercury into gold, a theoretical explanation of which process he based upon the firing of an electron into a mercury nucleus, the nuclear charge number being thus reduced by unity. This hypothesis, however, received a severe shock from Aston's subsequent observation that mercury does not possess an isotope of mass-number 197. Moreover atomic weight determinations allow of no distinction between "artificial" and "natural" gold. *Cf.* the discussion at the *Danziger Naturforschertagung*, 1925, published in the *Physikal. Zeitschr.*, 1925.

empirical proof of the accuracy of the modern conception, which regards all atoms as being composed of protons and electrons. On the other hand, the phenomenon of radioactive α-radiation appears to prove that the nuclei are *not directly* built up of protons and electrons, but rather that these components of the nuclei are grouped together especially frequently in the form of *aggregates* with a mass of four hydrogen atoms and a positive charge of two elementary quanta.

Light is thrown on the essential *rôle* played by *alpha-particles* in the construction of the nuclei, not only by the phenomena of radioactivity, but also by the fact that it is just the *most common atomic species* which have an *atomic weight divisible by four*. To elements with an atomic weight and atomic number of the form

$$A = 4n\,, \quad z = 2n \tag{1}$$

—and such elements are known for values of n from 3 to 10—we should, perhaps, ascribe a *pure alpha-structure* (*cf.* the first horizontal row in Table X). The ability of positive α-particles alone to combine into nuclei of this kind is explained by the fact that the α-particles themselves each consist of four protons and two electrons, and are thus able to attract one another electrostatically, when close enough and suitably arranged, despite their total positive charge.

Atomic weights divisible by four are represented, up to the end of the natural sequence, by comparatively common atomic species whose *atomic numbers* are *always even*. From the atomic weight 44 onwards, however (and to a certain extent, in consequence of isobares, even at 40), the *atomic weight* is *more than twice the atomic number*. If $4n$ be the atomic weight and $2n'$ the atomic number, the difference $(n - n')$ increases as we traverse the natural sequence, until it finally becomes equal to 13 (thorium has an atomic weight of $232 = 4 \times 58$ and an atomic number of $90 = 2 \times 45$).

This fact may be most readily explained on the hypothesis that the nuclei of these atoms also contain, besides positive α-particles of charge $(+\,2e)$, *neutralized alpha-particles* which, without affecting the nuclear charge, nevertheless contribute four units to the atomic weight. This hypothesis, due to *Miss Lise Meitner*, is principally supported by the experimental fact that, in the radioactive *disintegration series*, an α-rayer is frequently followed by two β-rayers, or a β-rayer by an α- and a β-rayer (*cf.* the previous Tables VII–IX in § 26).

A survey of the hypothetical construction of atomic nuclei from positive and neutralized α-particles is given in Table X. This refers to elements with even atomic numbers up to 42. Since the elements up to the atomic number 40 have already been investigated thoroughly for the existence of isotopes (*cf.* the earlier Table V in § 23), only the atomic species Mo_{96} is uncertain, although its existence is extremely probable, inasmuch as the " atomic weight " of molybdenum is an integer (96·0).[1]

TABLE X

Construction of Atomic Nuclei from α-Particles

(n α-particles, s of which are neutralized; below the chemical symbol are given the atomic number and atomic weight.)

$s = 0$ $A = 4n$ $z = 2n$	$n=3$ C 6, 12	$n=4$ O 8, 16	$n=5$ Ne 10, 20	$n=6$ Mg 12, 24	$n=7$ Si 14, 28	$n=8$ S 16, 32	$n=9$ A 18, 36	$n=10$ Ca 20, 40
$s = 1$ $A = 4n$ $z = 2n - 2$	$n=10$ A 18, 40	$n=11$ Ca 20, 44	$n=12$ Ti 22, 48	$n=13$ Cr 24, 52	$n=14$ Fe 26, 56	$n=15$ Ni 28, 60	$n=16$ Zn 30, 64	
$s = 2$ $A = 4n$ $z = 2n - 4$	$n=17$ Zn 30, 68	$n=18$ Ge 32, 72	$n=19$ Se 34, 76	$n=20$ Kr 36, 80				
$s = 3$ $A = 4n$ $z = 2n - 6$	$n=20$ Se 34, 80	$n=21$ Kr 36, 84	$n=22$ Sr 38, 88	$n=23$ Zr 40, 92	$n=24$ Mo ? 42, 96			

Besides the elements with an atomic weight divisible by four, those whose atomic weight is of the form $(4n + 3)$, where the *atomic number* is *always odd*, are of especially frequent occurrence. In particular, elements with an atomic weight and ordinal number of the form

$$(2) \qquad A = 4n + 3, \quad z = 2n + 1,$$

n having values of from 1 to 9 with the exception of 3, are represented by comparatively common atomic species.[2]

[1] Representatives of higher series than those shown in Table X would be, for instance, the isotopes of 50 (Sn), with atomic weights of 116, 120, and 124, s being put equal to 4, 5, 6. For isotopes of 54 (Xe), with atomic weights of 128, 132, and 136, $s = 5$, 6, or 7. For the isotope of 80 (Hg), with an atomic weight of 204, s is to be put equal to 11.

[2] These are: Li (3, 7), B (5, 11), F (9, 19), Na (11, 23), Al (13, 27), P (15, 31), Cl (17, 35), K (19, 39).

To many elements with an atomic weight divisible by four there corresponds an element with an atomic number greater by one, and an atomic weight higher by three. This indicates that the nuclei of such atoms perhaps contain, besides positive and neutralized α-particles, a single aggregate of mass 3 and charge $(+ e)$.

Atomic species with an atomic weight of $(4n + 1)$ or $(4n + 2)$ are not so common as those with an atomic weight of $4n$ or $(4n + 3)$. In this case also, the rule that the atomic number is even for an even atomic weight, and odd for an odd atomic weight, generally holds good; there are, however, exceptions which occur principally near the beginning of the natural sequence of the elements.[3]

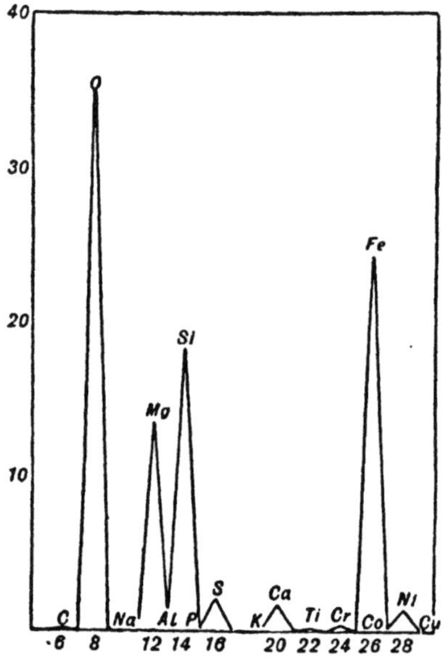

Fig. 36.—Relative proportions of elements in stone meteorites.

With respect to the number of isotopes, there exists a marked difference between elements of even and those of odd atomic number. Elements with *odd atomic numbers* have *few isotopes*. As the earlier Table V in § 23 shows us, they are either *pure elements*, or they consist of only *two isotopes*, the *atomic weights* of which are *odd* and *differ by two units*. (Apart from short-lived radio-elements, the third and fifth elements, lithium and boron, in which the difference in atomic weights is only unity, are exceptions.) On the other hand, elements with even atomic numbers generally have a larger number of isotopes, particularly for the higher atomic numbers.

Furthermore, it is found that elements of *even atomic number* are of *more frequent occurrence* than those of odd atomic number—obviously in consequence of a greater nuclear stability. This may be seen very clearly in Fig. 36,

[3] Such exceptions are, on the one hand, Li_6, B_{10}, N_{14}, and on the other hand, Be_9, Mg_{25}, Si_{29}.

which shows the relation between the average percentage by weight of the elements in stone meteorites and the atomic numbers (drawn as abscissæ). It can, moreover, scarcely be chance that the three gaps in the natural sequence of the elements all belong to odd atomic numbers.

As regards the *linear dimensions of the atomic nuclei*, we can draw some conclusions from the earlier-mentioned experiments concerning the *scattering of alpha-rays* in their passage through metal foils (*vide* § 21). For, according to Coulomb's Law, the scattering depends on the size of the nucleus as well as on its charge, so that the experimental results enable us to derive an *upper limit for the radius of the nucleus* (on the assumption that it is spherical in form). In this way a value of 4×10^{-12} cm. has been calculated as the upper limit for the radius of the *gold nucleus*, or less than a thousandth of the radius of the atom. Again, it follows from the formula giving the so-called *electromagnetic mass* of a spherical volume charge that, provided the mass of an electron is solely electromagnetic in origin, its radius must be about 2×10^{-13} cm., and that of a proton about 1×10^{-16} cm. If the mass of an electron were partly of another origin, the above values would represent lower limits, for, with a smaller radius, the electromagnetic mass of an electron would have to be larger than its actual mass. Thus the radius of the gold nucleus can be, at most, only about twenty times that of the electrons of which, together with protons, it is composed.

As a result of the small linear extent of a nucleus, its *internal energy* is, comparatively speaking, enormous. It is so large that the *equivalent mass* which may be ascribed to it, in accordance with the Principle of the Inertia of Energy (a consequence of the theory of relativity), may have a *noticeable influence on the atomic mass ;* this equivalent mass is obtained by dividing the energy by c^2. In order to grasp this, we have only to remember that even the kinetic energy of a *single* α-particle expelled from an atomic nucleus amounts [according to § 4 eqn. (2)] to $1{\cdot}3 \times 10^{-5}$ ergs; the equivalent mass is $1{\cdot}4 \times 10^{-26}$ grams, or almost 1% of the mass of a hydrogen atom.

If, now, the internal energy of a nucleus be W, a positive or negative sign being taken according as energy is used up or liberated in " packing together " the components of the nucleus, the mass will appear larger in amount by W/c^2, as compared with the sum of the masses of the components. When the nucleus contains n hydrogen nuclei (the mass of the electrons may be disregarded), the *atomic weight* must thus be given by

$$(3) \qquad A = n \times 1{\cdot}008 + \frac{W\,L}{c^2},$$

where L, as usual, denotes the Loschmidt number.

Conversely, from eqn. (3) we can calculate the internal energy. Let us take, for example, *helium*, for which A is found to be 4·00 and n is to be put equal to 4; then

$$(4) \qquad W = -\frac{0{\cdot}03 \times c^2}{L} = -\,4{\cdot}5 \times 10^{-5}\ \text{erg}.$$

The internal energy of an α-particle is therefore about *three times* as large as its kinetic energy for the largest velocity observed in α-rays.

It follows from the negative sign in eqn. (4) that, on *building up a helium nucleus* from its components, *i.e.*, on forming helium from hydrogen, *energy is liberated*. Since there are $1{\cdot}5 \times 10^{23}$ atoms in a gram of helium, $6{\cdot}8 \times 10^{18}$ ergs or, approximately, $1{\cdot}6 \times 10^{11}$ calories would be liberated on forming one gram. The evolution of heat connected with the synthesis of one gram of helium from one gram of hydrogen would therefore be some five million times as large as that associated with the combustion of one gram of hydrogen to form water. In their great internal energy lies the explanation of the fact that the α-particles themselves remain intact even when they disrupt other atomic nuclei.

The extraordinary stability of helium nuclei is probably the cause of the especial part they play in the building-up of more complex nuclei. Obviously the *synthesis of these higher nuclei* from helium nuclei is connected with a readjustment of energy, but, from the circumstance that the pure elements and isotopes have almost integral atomic weights (relative to O = 16), we must conclude that this energy change is much less than that accompanying the synthesis of helium. Indeed, the deviations of the atomic weights from integral values are so small that we cannot even say whether, for example, in the synthesis of an oxygen nucleus from four helium nuclei, energy is used up or liberated.

The energy required for a nuclear synthesis is partially liberated again in a radioactive disintegration, or in an artificial nuclear disruption. Since the transformation of an atom of uranium into an atom of uranium lead is associated with the expulsion of eight α-particles with a kinetic energy whose equivalent mass, in each instance, is nearly 1% that of a hydrogen atom, the difference in atomic weight between uranium and uranium lead must at any rate be greater than 32, and perhaps 32·06 or more. Actually it is more than 32, being between 32·1 and 32·2.

Thus, by the discovery of isotopes, and by drawing attention to the packing effect, recent atomic researches have succeeded in explaining the deviations of the atomic weights from whole-number values. Modern physics is hence justified in regarding the atomic weights as a clear proof that all elements are *composed of hydrogen*, just as *Prout* did more than a hundred years ago (in 1815), in the sense that, as we know now, the two constituents of the lightest of all atoms, the hydrogen nucleus and its satellite, go to build up the atoms of *all* the elements.[4]

[4] Recent researches on β-ray spectra seem to show that the structure of the nuclei is governed by *quantum relations*, just as is the arrangement of the electrons round the nucleus; *cf.* Ellis and Skinner, *Proc. Roy. Soc.*, *A* 105 (1924), pp. 185–198. It is also worthy of note that, if positive and negative electricity be interchanged in an atomic model constructed on a quantum basis, a model is obtained whose linear dimensions are a thousand times as small, so that it may be regarded as a model of a nucleus. The cause of this reduction lies in the difference in mass of protons and electrons. The conception of the nucleus as an "inverted" Rutherford atomic model is due to *Lenz*.

Chapter V

GENERAL THEORY OF SPECTRA AND OF ATOMIC STRUCTURE

§ 30. Röntgen Series.

In laying the foundations of Röntgen spectroscopy, *Moseley* discovered an important relation which permits an *interpretation of the lines in Röntgen spectra* from the point of view of the *quantum theory* and Bohr's theory of spectra. This relation expresses that the *frequency* of the strongest line in the K-series, the K_α-line, can be represented by the formula

$$\nu(K_\alpha) = \frac{3}{4} R (z - a)^2, \tag{1}$$

where R denotes the Rydberg constant, z the nuclear charge number of the element concerned, and a is a constant with a value differing but little from 1·6 for the majority of the elements.

In terms of Bohr's theory of spectra, eqn. (1) finds an obvious explanation if we consider the K_α-line as being due to a *transition of an electron from a two-quantum to a one-quantum orbit* [inasmuch as 3/4 is equal to $(1/1^2 - 1/2^2)$]. Provided, however, that we regard the atom, in rough approximation, as a hydrogen-analogue, the number a is to be interpreted as a so-called *screening constant.* The nucleus behaves as though a part of its charge, *viz.*, ae, were screened off ; for not only is the electron that undergoes the transition attracted by the atomic nucleus, but it is also *repelled* by the other electrons which, as well as it, surround the nucleus. Since, however, the screening constant of the K_α-line has almost the same value for the most various elements, this obviously means that the transition giving rise to the K_α-line must occur in a region of the atom which is *similarly constituted* for all atoms of arbitrary nuclear charge numbers (provided these are not too small). Apart from elements with very low nuclear charge numbers, the arrangement of the electrons moving in one- and two-quantum orbits must therefore be approximately the same in all elements.

For the strongest line in the L-series, the so-called L_α-line, Moseley likewise found the relation

$$(2) \qquad \nu(L_\alpha) = \frac{5}{36} R(z - b)^2,$$

where, again, b is a screening constant. Its value for most elements is about 3·5, but the variations in the values of the screening constant for the L_α-line are considerably greater than in the case of the K_α-line.

Eqn. (2) may evidently be interpreted as meaning that

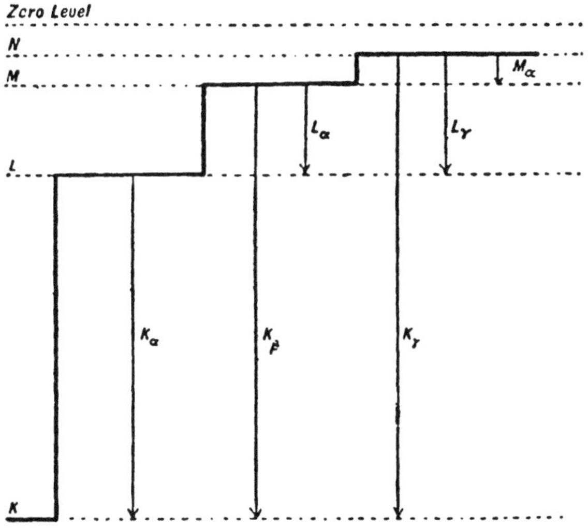

FIG. 37.—Diagrammatic scheme of the most important Röntgen lines.

the L_α-line arises from a transition of an electron *from a three-quantum to a two-quantum orbit.* The second strongest line in the K-series, generally known as the K_β-line, is due to a transition from a three-quantum to a one-quantum state, as follows from its wave-length. In like manner, the second strongest line in the L-series, usually denoted by L_γ,[1] arises from a transition from a four-quantum to a two-quantum state, and the strongest line in the M-series, the so-called M_α-line, from a transition from a four-quantum to a three-quantum state. The transition from the four-quantum to the one-quantum state gives rise to a weak line in the K-series, the so-called K_γ-line, whose wave-length is shorter than that of the K_β-line.

[1] The symbol L_β is used for another purpose.

A schematic representation of the origin of the above-named lines is given in Fig. 37, which is constructed according to the same principle as the earlier Fig. 5. The *one-quantum* level is called the *K-level*, the *two-quantum* the *L-level*, the *three-quantum* the *M-level*, the *four-quantum* the *N-level, etc.* The lines in the *K*-series are due to a transition to a *K*-level, those of the *L*-series to a transition to an *L*-level, and so on.

The diagrammatic representation leads at once to the following *combinations* giving the *frequencies* :—

$$\text{(3)} \qquad \begin{cases} K_\beta = K_\alpha + L_\alpha\,, \\ K_\gamma = K_\alpha + L_\gamma = K_\beta + M_\alpha\,, \\ L_\gamma = L_\alpha + M_\alpha\,. \end{cases}$$

These relations are found to be satisfied in great approximation ; the small deviations observed are to be explained by the fine structure of Röntgen spectra, which we shall discuss in the next section.

As we pointed out in § 9, every *energy-level* may be associated with a so-called *term* obtained on dividing the energy, taken with a positive sign, by the product *hc*. This term represents a *reciprocal wave-length*, and the wave-length of the line emitted in the transition from one level to another is then given by the relation

$$\text{(4)} \qquad \frac{1}{\lambda} = T' - T'',$$

where T' and T'' denote the terms corresponding to the initial and final levels. On replacing the terms by their reciprocal values, the *wave-lengths characterizing the levels*, eqn. (4) becomes

$$\text{(5)} \qquad \frac{1}{\lambda} = \frac{1}{\lambda''} - \frac{1}{\lambda'}\,.$$

Each energy-level may also be defined by an equivalent *potential*, expressed in *volts*, which is calculated from λ' or λ'' by means of Duane and Hunt's equation [§ 20 eqn. (3)].

Soon after the discovery of Röntgen spectra, experiment led to the important rule that the *excitation potential* necessary for the generation of a definite line is always *equal to the characteristic potential of the final level.* Hence, it is not the value of λ but always that of λ'' which determines the excitation potential in accordance with Duane and Hunt's equation. The converse also holds good. If the potential be gradually increased until the value corresponding to λ'' in Duane and Hunt's equation be reached, then at this potential *all the lines* whose *common final level* is defined by the wavelength λ'' will be *simultaneously excited.*

From this fact *Kossel*, in 1914, first drew the important conclusion that the *emission of a Röntgen line* can only occur when an electron in the emitting atom has been *previously* removed from its normal orbit to *beyond the limits of the atom.* Thus, a *place in the atom* must first, as it were, *fall vacant.* If this vacant place corresponds to a quantity of energy $(-h\nu'')$, and if it be filled by an electron which was previously situated at an energy-level $(-h\nu')$, a line of frequency $(\nu''-\nu')$ will appear. For the generation of this line, however, the energy $h\nu''$ must have been *previously expended.*

A convincing proof of the accuracy of Kossel's conception is afforded by the appearance of *absorption band heads in continuous Röntgen spectra.* If the *Röntgen line spectrum* of an element be *excited by primary continuous Röntgen radiation*, instead of by cathode rays, the *wave-lengths characteristic of the energy-levels in the line spectrum excited* manifest themselves as heads of absorption bands in the primary Röntgen spectrum.

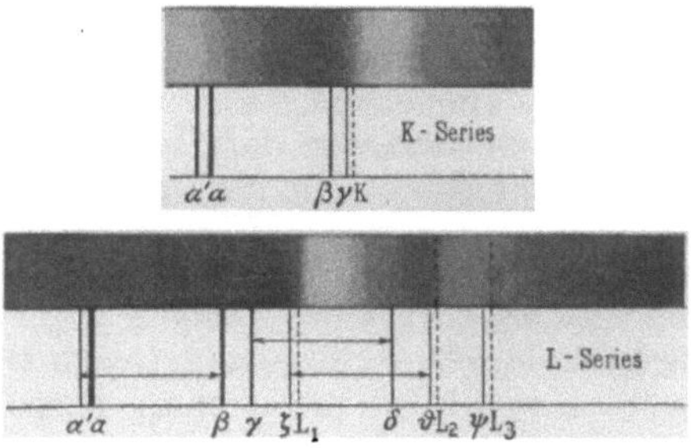

FIG. 38.—Absorption band heads in the continuous Röntgen spectrum.

The explanation of this phenomenon follows immediately from Kossel's conception. For, in accordance with Einstein's law of light-quanta, we must suppose *light-quanta* to be *absorbed from the primary continuous radiation* and to be *transformed into light-quanta of the secondary line radiation.* Hence, according to Kossel's theory, a line in the K-series can be excited only by primary rays of frequency *at least* as great as the characteristic frequency of the K-level. On the other hand, in the excitation of K-lines, no primary rays are absorbed, the wave-lengths of which are larger than the wave-length characterizing the K-level. Thus, we may expect to find the head of an absorption band at this wavelength, and likewise at the wave-lengths characteristic of the L- and M-levels.

These expectations are borne out by actual experiment.

In the upper photograph reproduced in Fig. 38, we see the absorption band head of the K-series in silver; beneath it are inserted the four lines of the corresponding K-series (their wave-lengths being larger). The primary Röntgen radiation was generated in a 40 kilovolt tube, so that, by Duane and Hunt's equation, the continuous Röntgen spectrum extended down to about 300 X.U. The sharp head to the absorption band, seen in the photograph, corresponds to a wave-length of 485 X.U. Since the wave-lengths of the absorption band heads are characteristic of the energy-levels themselves, thus corresponding to the difference between a level and the zero-line, the *heads of the absorption bands* also represent *series limits*, as is likewise seen from Fig. 38.

The problem of experimental Röntgen spectroscopy consists chiefly in the *exact determinations of wave-lengths*, both of absorption band heads and of individual emission lines. Much the most accurate method of measurement is the *direct* one based on *Bragg's equation of crystal reflection*, but this method can only be applied within a limited region of wave-lengths. The majority of wave-lengths are measured by means of a *rock-salt* or *calcite crystal*, which, as already mentioned in an earlier section (§ 19), can be used for wave-lengths up to about 5,000 X.U. Wave-lengths more than twice as large can be reached with the help of *selenite* or *mica*, and as much as more than three times as large by the use of crystals of organic compounds. Exact measurements by means of crystals have been perfected, principally by *Siegbahn*, to such a degree that, at the present time, Röntgen wave-lengths can be determined with an accuracy of about a millionth part of the measured value. (That is to say, wave-lengths can be compared with one another, or referred to an idealized lattice constant with that accuracy.) [2]

In comparison with these direct precision methods, only a secondary value can be attached to the *indirect methods*. These are, however, of the greatest importance in the *boundary region* between the largest wave-lengths measurable with crystals and the shortest wave-lengths in the ultra-violet. The most important of these indirect methods is based on *Duane and Hunt's equation*, the wave-length of the *series limits* being thereby derived from the measured excitation potentials (defined by the values at which absorption suddenly commences).

In exceptional cases, the characteristic wave-lengths can also be measured with an *optical diffraction grating* when the

[2] *Vide* § 19.

nuclear charge numbers are *small*, because, with decreasing nuclear charge number, all series gradually shift over to the ultra-violet region.

As regards the *K-series*, the *series limit* is known from measurements of the *absorption band heads* for the elements between 12 (Mg) and 92 (U); for Mg it amounts to 9511·2 X.U., and for U to 107·5 X.U. The wave-lengths of the individual *lines* in the K-series are known for the elements from 11 (Na) to 60 (Nd) and, in addition, for 74 (W), 77 (Ir),

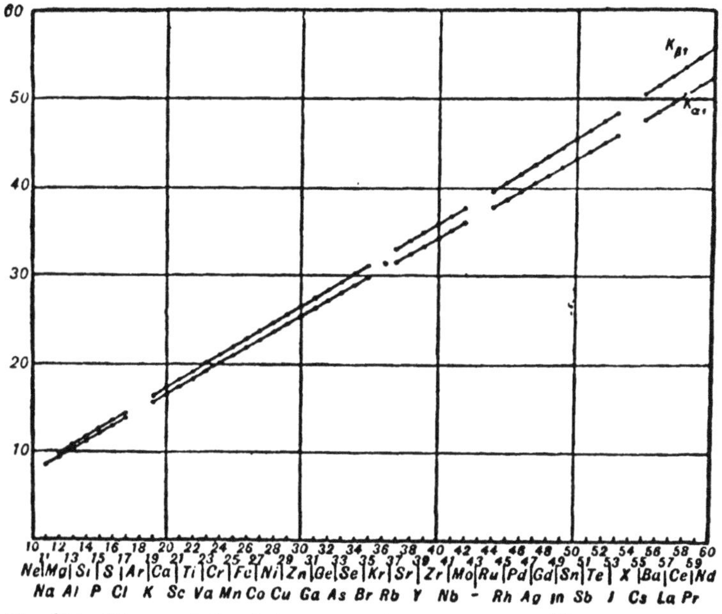

FIG. 39.—Linear relation between atomic number and $\sqrt{\nu}$ for the K_α-line.

and 78 (Pt). The extraordinary accuracy with which the *linear relation* between the square-root of the frequency of the K_α-line and the atomic number is satisfied is evident from Fig. 39.

The K_α-line for Na lies as high as 12,000 X.U.; for elements of still lower ordinal number, it lies in the boundary region between Röntgen rays and ultra-violet rays. By means of Duane and Hunt's equation, however, the K-limit has been determined in this boundary region also for 8 (O), 6 (C), 5 (B), 4 (Be), and 3 (Li); *e.g.*, for C it lies at 47, and for Li at 290 Å.U.

The limit (or, as we should rather say, having regard to the fine structure to be discussed subsequently, the limits) of the *L-series* is known for the elements between 55 (Cs) and 92 (U). The emission lines of the *L*-series are known from 26 (Fe) to 92 (U). For Fe the wave-lengths of the *L*-lines are as large as 17,000 X.U.; for U they are about 1,000 X.U.

In 1921, *Millikan*, by means of his *high-vacuum spectrograph*, was able to observe and measure the *L*-series in the *extreme ultra-violet* for 13 (Al), 12 (Mg), and 11 (Na). One wave-length for Al amounts to 136·6 Å.U.; this also represents the shortest wave-length yet measured in the ultraviolet. Indirect measurements have enabled the *L-limit* to be determined also for 3 (Li), 4 (Be), 5 (B), 6 (C), 8 (O), and 9 (F).[3]

At the present time, the absorption band head in the *M*-series (discovered by *Siegbahn*) has been directly measured only in the cases of three elements, *viz.*, 83 (Bi), 90 (Th), and 92 (U). Wave-length determinations principally carried out by *Stenström* have been made for the elements between 66 (Dy) and 92 (U). It has also been possible to measure the *M*-limit indirectly for several elements, *e.g.*, for 42 (Mo), in which case it lies at about 34 Å.U.

The *N-series* was discovered by *Dolejsek* in 1922;[4] up till now measurements have only been made for 83 (Bi), 90 (Th), and 92 (U). The *N*-series is so soft that its lines, even for these highest elements, lie right on the boundary of the wave-lengths which can be determined by means of crystals.[5]

§ 31. The Fine Structure of Röntgen Spectra.

On closer investigation of Röntgen spectra, it was soon found that the number of lines observed was too large for the simple assumption of a single *L*-level, a single *M*-level, *etc.*, to suffice for their explanation. It appeared necessary, in fact, to distinguish between *various modifications of the levels* (except for the *K*-level). This necessity was also rendered evident by researches on *absorption band heads*, which showed that the *L*-limit is *threefold* and the *M*-limit *fivefold* in those

[3] *Cf.* the summary by MacLennan in *Nature*, **113** (1924), p. 217.

[4] Dolejsek, *Zeitschr. f. Phys.*, **10** (1922), p. 129.

[5] In 1924 Wentzel closely investigated the so-called "spark lines" in the Röntgen spectra, which are produced when more than one electron is removed from the interior of an atom. *Cf.* Sommerfeld's "*Atomic Structure and Spectral Lines*," 4th edn., ch. iv., § 9.

elements where they are observed. Only the K-limit appears to be single in every case (*vide* the earlier Fig. 38).

As soon as the numerous lines in the different series had been determined, it became one of the most important problems in Röntgen spectroscopy to refer the manifold of these lines, through their combinational relations, to a less complicated multiplicity of energy-levels, and thus to create a *scheme for the origin of Röntgen lines.* We owe the final solution of this difficult problem, obtained in 1922, principally to four investigators: *Smekal, Wentzel, Coster,* and *Bohr.*

According to Bohr, Landé, and Sommerfeld, this scheme is most simply stated on the following lines. *Each level* is characterized by the corresponding letters K, L, M, N, O, P, Q, where K is associated with the principal quantum number 1, L with the principal quantum number 2, *etc.* Furthermore, we assign to each level two indices, the first of which represents the *subordinate quantum number* (in the sense of § 12), and the second the so-called *inner quantum number.* Thus we may write the general symbol S_{kj}, in which S represents K, L, M, N, *etc.*, whilst k is equal to 1, 2, or any integer less than an upper limit depending on the nuclear charge number of the atom, but in no case exceeding the principal quantum number. Finally, the inner quantum number j either may be equal to k, or it may have a value less by one, a lower level corresponding to the smaller value of j. A level for which $j = k$ is conventionally described as a *normal* level, whilst when $j = k - 1$, we speak of an *anomalous* level. When $k = 1$, only the normal level can exist.

The largest number of possible levels associated with a principal number n is thus $(2n - 1)$, though this number is not necessarily realized in the atoms of all elements. For example, the five possible M-levels are thus represented in ascending order by the symbols

$$M_{11},\ M_{21},\ M_{22},\ M_{32},\ M_{33}.$$

As stated by *Bohr*, the *selection principle of Röntgen lines* may be expressed in the following form: The only permissible transitions between the various energy-levels are those in which the *subordinate number changes by unity,* whilst the *inner quantum number either remains unaltered or likewise changes by unity,* and these transitions lead to the generation of Röntgen lines. Hence the two conditions

$$(1) \qquad \begin{cases} \Delta k = \pm 1, \\ \Delta j = \begin{cases} 0 \\ \pm 1 \end{cases} \end{cases}$$

must be satisfied. (It is taken for granted that, in such a transition, the principal quantum number n changes.)

In Fig. 40 is reproduced the *scheme of Röntgen lines* as determined by the above-stated selection principle for *elements of the highest atomic numbers* (from emanation onwards). With these highest elements there are 1 K-level, 3 L-levels, 5 M-levels, 7 N-levels, 5 O-levels, and 3 P-levels.[1]

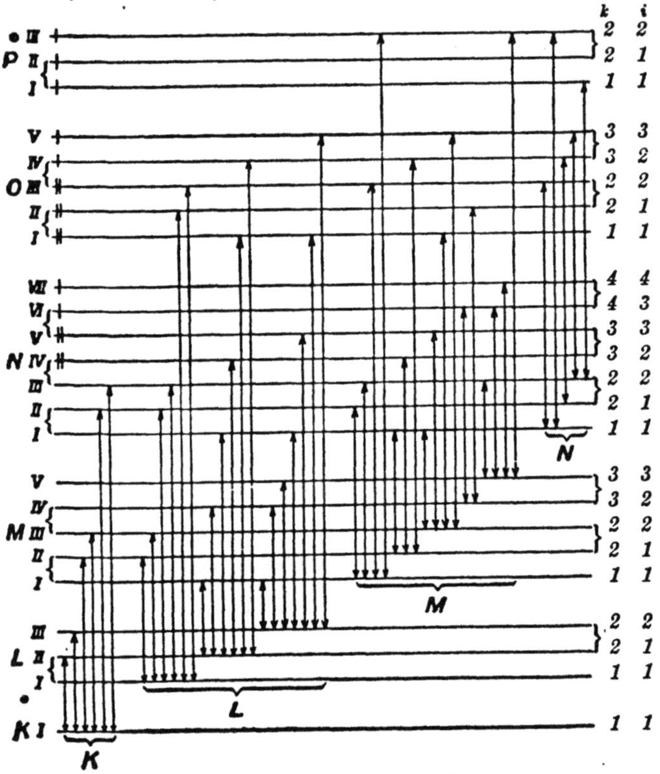

FIG. 40.—Scheme of Röntgen lines.

We see from Fig. 40 that the transition from an L- to a K-level, for instance, is associated with two lines, while seven lines correspond to the transition from an M- to an L-level.[2]

[1] In the O- and P-levels the largest conceivable number (9 and 11 respectively) is not attained; but it is in the K-, L-, M-, and N-levels.

[2] Besides the lines shown in the scheme, there are, of course, other weaker lines, which arise when more than one electron is removed from the interior of the atom. For instance, with each of the elements Mg, Al, Si, six different lines, due to a transition from a two-quantum to a one-quantum state, have been observed.

It is also evident from Fig. 40 that the individual groups of lines are resolved into *sub-groups* distinguished by the same final level, and thus by the same absorption band heads. In the figure single vertical dashes denote those levels which gradually lapse as we descend the natural sequence of the elements from emanation to xenon. In the case of the elements near krypton, other levels are also lacking, these being denoted in the figure by double vertical dashes. An example of a photograph of an L-series is to be found in Fig. 41, which gives the L-spectrum of rhodium, containing 11 lines.[3]

The individual *energy-levels* or, what amounts to the same thing, the individual *terms* are known at the present time with fair accuracy for most elements. They were determined either directly by measurements of absorption band heads or, with the help of the above-given scheme of lines, by measurements of the emission lines. The relations between the terms and the atomic numbers are shown in Fig. 42, where the ordinates represent the *square roots* of the terms referred to the Rydberg term as unit.[4] Only the portions of the curves drawn in full, however, correspond to *actually measured* values ; the numerical symbols and lines added underneath the figure will be considered later (§ 35).

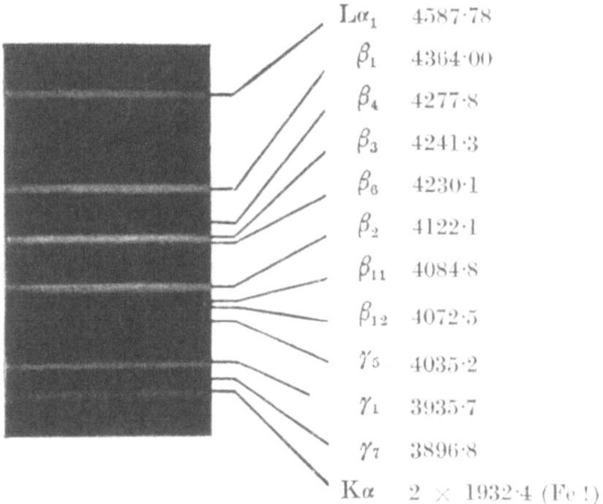

FIG. 41.—L-series of rhodium.

[3] The lines have been marked with the corresponding wave-lengths in X.U. As well as the rhodium lines, the K_α-line of iron is also to be seen in Fig. 41, iron being present as a small impurity in the rhodium. In Röntgen spectroscopy it is possible to detect impurities amounting to only from a hundredth to a tenth of 1 *per cent.*

[4] The terms, which are reciprocal to the wave-lengths, are therefore obtained on multiplying the squares of the ordinates by 109,737. The P-terms, which are still rather uncertain and have, of course, only been calculated for the highest of the elements, have been omitted in Fig. 42.

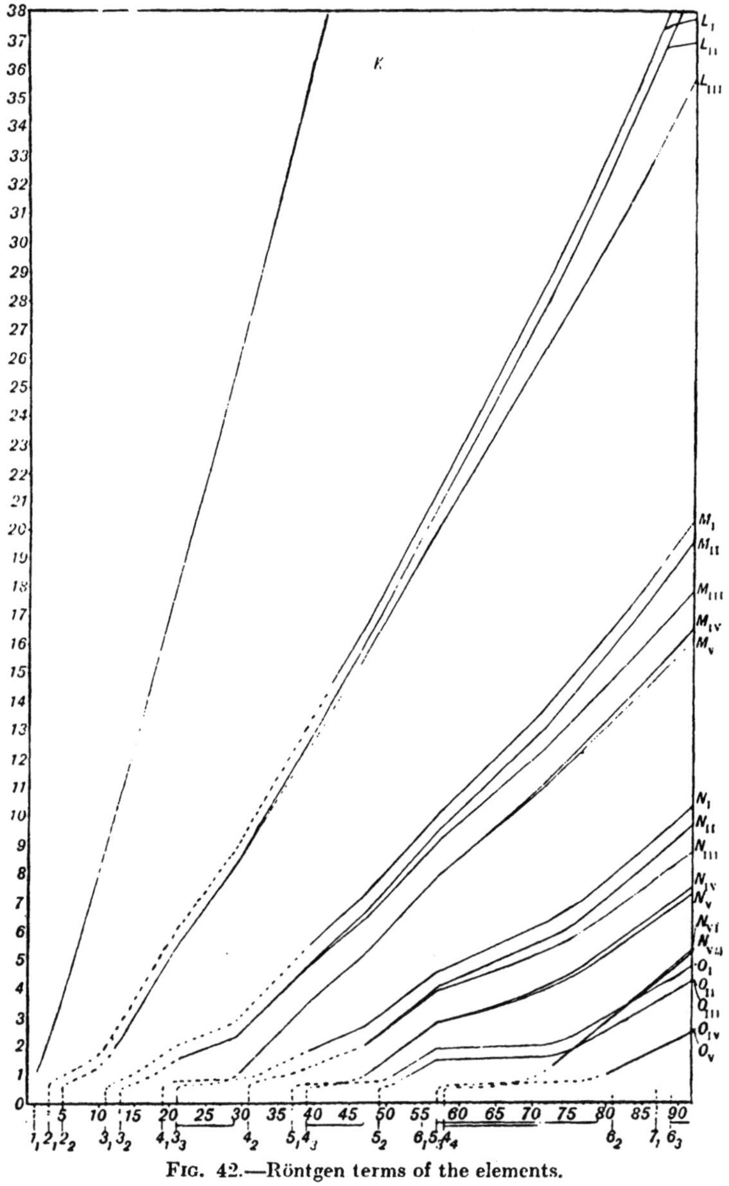

FIG. 42.—Röntgen terms of the elements.

The course of the various curves is more or less linear, in accordance with Moseley's Law; the larger the principal quantum number, the smaller is the angle between the curve and the axis of abscissæ.

Within every group characterized by one and the same value of the principal number, each pair of adjacent levels forms a so-called *level doublet.* There are, however, two kinds of doublets to be distinguished. The first is formed by two neighbouring levels with the same principal number and the same subordinate number; in this case we speak of a *regular doublet.* The second kind of doublet is constituted by two adjacent levels for which the values of n and j are the same, while the values of the subordinate number k differ by unity; we then speak of an *irregular doublet.* Within each group regular and irregular doublets alternate with one another. In the earlier Fig. 40, the regular doublets are marked by brackets on the right-hand side, and the irregular doublets by brackets on the left-hand side. The difference between the energy values in a doublet, divided by the elementary quantum of action, is known as its *frequency difference.*

Now, as *Sommerfeld* discovered in 1916, the frequency difference of a *regular* doublet is given by the simple relation

$$\Delta\nu = \Delta\nu_{H}\,(z - b)^4\,, \tag{2}$$

where $\Delta\nu_H$ denotes the so-called frequency difference of the hydrogen doublet [§ 13 eqn. (11)], z the atomic number, and b the relevant screening constant.[5]

Wave-length difference and frequency difference are connected by the simple relation

$$\Delta\lambda = -\frac{c}{\nu^2}\,\Delta\nu\,, \tag{3}$$

which is at once obtained by differentiating the equation

$$c = \lambda\nu\,.$$

Since, by Moseley's Law, the square root of the frequency is proportional to the atomic number less the screening constant, it follows from eqns. (2) and (3) that the *wave-length difference of a regular doublet* is almost *independent* of the *atomic number.*

A simple rule, discovered by *G. Hertz* in 1920, likewise holds good for *irregular doublets:* the *frequency difference* of irregular doublets proves to be almost *independent of the*

[5] *E.g.*, for the doublet of the energy-levels $L\,(2, 2)$ and $L\,(2, 1)$, b is equal to 3·5.

atomic number. Hence, in Fig. 42, the curves of two terms constituting an *irregular doublet* have a nearly *parallel* course, whereas the curves of two terms forming a *regular doublet diverge* more and more as the atomic number increases.

In conclusion, we may just mention how the *wave-lengths* of the *γ-rays of a radioelement* can be derived from its *magnetic β-ray spectrum*, provided the *energy-levels* of an element *isotopic* with it are known from Röntgen-spectroscopic measurements. According to the fundamental relation obtained in an earlier section (§ 5), the kinetic energy of a secondary β-particle is given by the equation

(4) $$E_{kin} = h\nu + W,$$

where ν is the frequency of the primary γ-rays and W the work necessary to remove the β-particle from its usual orbit "to infinity."

We shall now assume that the magnetic β-ray spectrum of a radioelement contains two lines corresponding to velocities v' and v''. The difference in these values is obviously explained by the fact that the β-particle is *liberated from a different level* in either case. If, however, the liberation is effected in each instance by γ-rays of the *same frequency*, we find, on applying eqn. (4) to each case and subtracting,

(5) $$E_{kin}' - E_{kin}'' = W' - W'';$$

in calculating the kinetic energy, the dependency of the mass on the velocity must be taken into account [according to § 3 eqn. (6)].

Now if both secondary β-rays really were generated by primary γ-rays of the same frequency, we see that the energy difference $(W' - W'')$ divided by h gives the frequency of a known *Röntgen line* of the isotopic element. This being the case, however, we can immediately find out from what levels the secondary β-particle was liberated in the two instances, provided we know the energy-levels of the isotopic element. We have only to replace W in eqn. (4) by the energy of the level in question (*i.e.*, by the spectral term multiplied by hc), when, since the kinetic energy of the secondary β-particle is known, the frequency of the primary γ-ray can at once be calculated from eqn. (4).

By means of this method, devised by *Ellis* and by *Miss Lise Meitner*,[6] it has been found, for example, that two lines in the β-ray spectrum of *radiothorium*, corresponding to

[6] C. D. Ellis, *Proc. Roy. Soc.*, *A* 99 (1921), p. 261; L. Meitner, *Zeitschr. f. Phys.*, 9 (1922), pp. 131 and 145.

velocities of 51 and 47% of the velocity of light, are associated with values of the work of liberation, which are identical with the energy values defining the M_I- and L_I-levels in the isotope, thorium; it follows from this that the wave-length of the primary γ-rays is 14·6 X.U. To mention a second example, three lines are found in the magnetic spectrum of thorium B which are associated with velocities of 63·0, 71·4, and 72·1% of the velocity of light. These lines correspond to the values of the work of liberation from the K-, the L_{II}-, and the L_I- levels of the isotope, lead; and it follows that the wave-length of the primary γ-radiation has a value of 52 X.U.

§ 32. Excitation and Ionization of Atoms.

From the spectrum of the simplest of all atoms, hydrogen, it is clear that an electron revolving around the atomic nucleus can leave its usual orbit and revolve *temporarily in an orbit of higher quantum number.* Quite generally, we say that an atom is *excited* when one of its electrons describes an orbit *other* than its normal one, the electron then having a *higher energy* than usual. The *excitation* of an atom thus always demands an *introduction of energy*, while, on the *return of the atom to its normal state, energy is liberated.*

As subsequent spectroscopic considerations will show, the multiplicity of higher levels, to which the electron can temporarily be raised, is very great. Each of these levels, however, which are not present in the normal atom and can only be realized when the atom is excited, is associated with a quite definite energy value, just as is the normal state. Hence, from amongst the manifold of excited states there must be one state that is distinguished by its having the *smallest excess of energy over the normal energy value.* This state is termed, in a narrower sense of the word, simply the excited state, and its surplus energy is called the *excitation energy* of the atom. It therefore represents the smallest quantity of energy which a normal atom of the element concerned can ever take up in connection with a change of its internal energy.

The excitation of atoms is, of course, most readily effected with the *most loosely bound* electrons. These are obviously the *peripheral electrons*, which an atom loses when it enters into chemical combination as a positive ion, and which are therefore called the *valency electrons.* Since the elements in the first group of the periodic system are positively univalent, those in the second vertical column positively divalent, and so on, we must assume that the atoms of the elements in the first group each possess a single valency electron, those in the second group two valency electrons each, and so on.

In general, changes in the orbit of the peripheral valency electron are probably responsible for those variations of the internal atomic energy which, in accordance with Bohr's *frequency condition*, manifest themselves in the *optical spectra*. For, in the optical region, a light-quantum is roughly a thousand times smaller than in the region of Röntgen rays, so that optical lines can only be due to those electrons for which, because they are loosely bound, a change of orbit can be effected by a relatively small expenditure of energy. Moreover, it is easy to understand why the excitation of an atom should, in general, find expression only in the peripheral electrons, because it is only these electrons that, since they describe the highest-quantum orbits in the normal atom, will, on a transition to an orbit of still higher quantum number, find it unoccupied. An inner electron, on the other hand, must, on a transition to a higher-quantum orbit, generally find it already completely occupied (as follows from concepts of the grouping of electrons, to be discussed in greater detail later).

We must assume that an excited state of the atom is not only abnormal, but also extremely *unstable*. After a very short sojourn, which we shall later find to be of the order of 10^{-8} sec., the *return to the normal state* takes place. Furthermore, provided the surplus energy is not changed into kinetic energy, this return is accompanied by the *emission of a light-quantum* whose frequency, according to Bohr's condition, is equal to the quotient of the energy released and the elementary quantum of action. If the atom was not in the lowest of the excited states, but in one more highly excited, the return to the normal state may, of course, follow *by steps*, so that the emission spectrum then produced will contain lines for which both the initial and the final levels are possible only in an excited state.

Conversely, the excitation of an atom, or its transformation from an already excited to a more highly excited state, can also be effected by the *absorption of a light-quantum*, provided the product of the frequency of the light-quantum and the elementary quantum of action be equal to the difference in energy between the actually existing and a higher-quantum state.

A far more frequent way of exciting atoms than by

absorption, however, is by supplying the requisite energy from the *heat store* of the body concerned. According to the fundamental concept of the kinetic theory of heat, incessant *collisions* occur in a gas between the rapidly moving molecules. By virtue of these collisions, *energy of translation* can of course be *changed into intra-atomic energy*, but when the atom struck was previously in the normal state, this can only happen if the relative kinetic energy is at least as large as the excitation energy. (In the monatomic vapours of the metals in the first two groups, which have been chiefly investigated, as well as in the rare gases, the concepts of atom and molecule are synonymous.) Now, when atoms thus excited by collisions return to the normal state with the emission of light-quanta, the well-known phenomenon of *thermo-luminescence* is observed. Moreover, the atoms of a gas can be excited, not only by sufficient heating, but also by collisions with rapidly moving ions *in electrical fields*, as is shown by the *luminescence of rarefied gases in Geissler tubes*.

The *excitation energy* in the narrower sense of the word, *i.e.*, the smallest quantity of energy which a previously normal atom can take up, may at once be derived from optical spectra, and also by other means, as we shall show in detail in the following sections. For the alkali metals its value lies between 2×10^{-12} and 3×10^{-12} ergs, being smallest for cæsium. For the elements of the second group in the periodic system, the excitation energy is two to three times as large as for cæsium ; for lead (in the fourth group) it is a little smaller than for cæsium, and for thallium (in the third group) it is about 20% smaller. The spectral lines corresponding to excitation in the narrower sense of the word lie in the optical region for most of the alkali metals and for Mg, Ca, Sr, and Ba, whereas for Zn, Cd, and Hg they fall in the ultra-violet.

The *mean energy* possessed by a monatomic *gas molecule* at an absolute temperature T, in virtue of its *translation*, is, as follows from the kinetic theory of gases, equal to $3/2\, k\, T$, where k, the so-called Boltzmann constant, is equal to $1{\cdot}372 \times 10^{-16}$ ergs/deg.[1] At room temperature, for which T is about 300, the mean value of the translational energy of a monatomic molecule is hence about 6×10^{-14} ergs. Furthermore, according to a fundamental relation of molecular statistics, the *fraction* (α) of molecules whose trans-

[1] In the case of molecules of diatomic gases, the factor 3/2 is replaced by the factor 5/2. Boltzmann's constant is the quotient of the gas constant and the Loschmidt number.

lational energy is greater than x times the mean value is, quite generally, equal to e^{-x}.

Thus, for cæsium at room temperature, the fraction (α) of the atoms whose translational energy exceeds the excitational energy is about 10^{-13} (for x is approximately equal to 30). At a temperature of 1,200° C. (= 1,500° abs.) x becomes roughly equal to 6, and hence α to about 1/400. Whereas, therefore, at room temperature the fraction of atoms possessing a kinetic energy capable of exciting another atom by collision is entirely insignificant,[2] at 1,200° C. this fraction is large enough to produce a noticeable luminescence in the vapour.

Theory teaches us to expect that, when any vapour commences to glow with rising temperature, the first spectral line to be observed will be that corresponding to the transition from the lowest excited state to the normal state. The longer the wave-length of this line, the smaller will be the excitation energy, and the lower the temperature at which emission of the line will begin. As this temperature is exceeded, we must further expect that the transfer of translational energy will be sufficient to produce higher and higher excitations, so that new spectral lines will be continually appearing. In point of fact, all these theoretical conclusions are confirmed by experiment.

The energy necessary to remove a valency electron completely from its atom can also be derived from the optical spectrum (and by other methods which we shall discuss later). For, when the lines in the optical spectrum can be referred to terms, we have only to multiply by hc the term corresponding to the lowest level, *i.e.*, the term with the largest numerical value (known as the *fundamental term*), in order to obtain this energy, the so-called *ionization energy*, in ergs.

For the alkali metals, and also for elements in the second group of the periodic system, the ionization energy proves to be two or three times as large as the excitation energy. Hence, if we now refer the fraction α to the process of ionization instead of to that of excitation, as previously, we should expect to find the same value of α for an absolute temperature two or three times as high. Actually, however, ionization is to be observed for much smaller values of α than is excitation. For, on introducing the salts of a metal into a

[2] Actually, it is not the mean translational energy, but the mean value of the relative energy on collision, that comes in question. It follows from the kinetic theory of gases, however, that the two quantities only differ by an inconsiderable amount.

flame, the smallest ionization can at once be observed by a noticeable *electrical conductivity of the flame*, for the detection of which we possess in the galvanometer an extraordinarily sensitive instrument. As theory leads us to expect, the flame conductivity at a given temperature is greater the smaller the optical fundamental term of the metal concerned.

Ionized atoms can, of course, be excited in an entirely similar way to neutral atoms, by the removal of one of the remaining valency electrons from its normal orbit to a higher-quantum orbit. Thus, the elements may give, besides the spectra of their neutral atoms, *spectra of their ionized atoms*, the simplest example of which is afforded by the spectrum of ionized *helium*, described in an earlier section (§ 10). The spectra of ionized atoms obviously presuppose higher temperatures or stronger electrical fields than the spectra of neutral atoms. Whereas the spectra of neutral atoms are generated, apart from in Geissler tubes, principally in the electric arc, the spectra of ionized atoms are best produced by means of electric spark discharges. The two varieties of spectra are therefore commonly distinguished, from their modes of production, as *arc spectra* and *spark spectra*. According as we are dealing with single, double, or triple ionization, we speak broadly of a simple spark spectrum or of a spark spectrum of the second or third order respectively.

It is evident that the spark spectrum of an element of atomic number $(z + 1)$ must exhibit a strong resemblance to the arc spectrum of the element of atomic number z, except that, when the spectrum is approximately that of a "hydrogen analogue," the *terms* will appear *multiplied by four*, as was indeed the case with the simple example of ionized helium. Similarly, in spark spectra of the second and third orders, the terms appear multiplied by nine and by sixteen respectively (*cf.* the following section, § 33). The so-called *spectroscopic displacement law*, discovered by *Sommerfeld* and *Kossel* in 1919, holds quite generally. According to this law, the spark spectra of the elements in any group of the periodic system resemble in character the arc spectra of the elements in the previous group, especially as regards their fine structure (which we shall discuss later).

The easiest spark spectra to produce are those of the second group elements, a temperature of 2,000° C. being sufficient for this purpose, for their simple ions still contain one loosely bound electron. Arc and spark spectra of these elements are thus often to be observed together. On the other hand, it is extremely difficult to generate the spark spectra of the alkali metals, since their atoms in the neutral state only contain a single valency electron.

These conclusions from Bohr's theory of the atom provide an explanation of several apparently paradoxical results in *solar spectroscopy*, as was first pointed out in 1920 by the Indian physicist, *Saha.*[3] It had previously been quite inexplicable why sodium should appear very strongly in the solar spectrum, whereas potassium appears only weakly, and rubidium and cæsium are entirely absent. According to Saha's striking explanation, the alkali metals are present in the sun, just as in the earth. *To some extent*, however, they *remain unnoticeable*, because, in consequence of the high temperature, they are almost invariably *ionized*, and their spark spectra, the lines of which lie in the ultra-violet, escape spectroscopic observation. That the arc lines of Na occur strongly in the solar spectrum, those of K only weakly, and those of Rb and Cs not at all, follows from the fact that Rb and Cs have the smallest ionization energies, K having a somewhat higher, and Na a considerably larger value.[4] In contrast to the alkali metals, the alkaline earths are represented in the visible portion of the solar spectrum by the lines of their ionized atoms as well; for instance, the strong Fraunhofer lines *H* and *K* both belong to the spark spectrum of calcium.[5]

Moreover, Saha found in Bohr's theory a simple explanation for the remarkable fact that the heavier element calcium is to be observed in the spectrum of the sun's atmosphere at much greater heights than the much lighter element hydrogen. For the calcium lines lying in the visible region of wavelengths require a smaller excitation energy than the hydrogen lines lying in the same region; the hydrogen lines which are *readily* excited belong to the ultra-violet.[6] In the higher regions of the sun's atmosphere, where the temperature is less than in the lower regions, the energy transferred by molecular collisions therefore suffices to excite the visible calcium lines, but not the visible hydrogen lines.

[3] *Phil. Mag.*, **40** (1920), pp. 472 and 809, and **41** (1921), p. 267.

[4] *Cf.* the later Table XVII, in § 37.—Besides the differences in ionization energy, the abundance or rarity of the elements, it is true, also play a part.

[5] The earlier view that calcium is the most abundant element in the sun's atmosphere has, of course, been proved erroneous by Saha's researches.

[6] The hydrogen line corresponding to the excitation energy, in the narrower sense of the word, is the first line in the Lyman series (1,216 Å.U.).

§ 33. Optical Spectra.

The optical spectra of the elements can be produced both by the emission and by the absorption of light-quanta. As was pointed out in the previous section, *absorption spectra* arise from the transition of an electron into a higher-quantum level, either from the normal orbit or from an orbit due to previous excitation. *Emission spectra* are produced by the return of the atom from an unstable to the normal state, whether this return takes place in one or in several stages.

Researches on optical spectra commenced with the observation by *Fraunhofer*, in 1814, of the absorption lines in the solar spectrum. In 1860 *Kirchhoff* and *Bunsen* founded *spectrum analysis* with their discovery of emission spectra. They proved that the bright lines in these spectra are characteristic of the elements, and the constant wave-lengths entirely independent of the temperature and chemical combination. Furthermore, they were able to demonstrate the coincidence of the Fraunhofer lines with the emission lines of certain elements, and to give a theoretical explanation of the former. The possibility of the existence of simple numerical relations between the wave-lengths of the lines in a spectrum was first pointed out by *Balmer* in 1885, in the particularly simple instance of the hydrogen spectrum (as has been discussed at length in an earlier section). About 1890, *Kayser* and *Runge* discovered simple series relations also in the spectra of the *alkali metals ;* and, at about the same time, *Rydberg* succeeded in expressing these relations quantitatively by formulæ which exhibited a strong resemblance to Balmer's formula, and in which a fundamental part was played by the so-called Rydberg constant, already known from the hydrogen spectrum. Series relations similar to those for the alkali metals, for which they are most clearly marked, were subsequently established for many other elements as well.

The common mark of distinction of all optical *series* is (as has already been shown by the simple example of the optical hydrogen series) the *decreasing intensity* with *decreasing wave-length*, together with the *convergence of the lines*, also with decreasing definition, towards a *limit* corresponding to the final level in the case of emission.

In those elements for which an arrangement of the optical spectral lines into series was possible, the most important series were soon recognized to be three in number, and these were distinguished as the *principal series*, the *first* or *diffuse subordinate series*, and the *second* or *sharp subordinate series*. The two subordinate series, which are clearly distinguished by the different sharpness of their lines, are found to have a *common limit* which, for the elements in the first two groups of the periodic system, always has a greater wave-length than the limit of the principal series. Subsequently, a fourth series of comparatively large wave-lengths was added to the three above-named series, and this is known as the *fundamental series* or, after one of its discoverers, as the *Bergmann series*.[1]

Later considerations have shown that the frequencies of the most important lines in the series named can be represented as differences between the members of *four series of terms*; these are distinguished by the initial letters of the words "sharp," "principal," "diffuse," and "fundamental" as s-, p-, d-, and f-terms.[2] The above four series arise on emission from the transitions represented in the following scheme :—

(1)

Principal series	. . . . .	$p \rightarrow s$,
1st (diffuse) subordinate series	.	$d \rightarrow p$,
2nd (sharp) subordinate series	.	$s \rightarrow p$,
Bergmann series	. . . . .	$f \rightarrow d$.

Investigations on the separability of the terms in the Zeeman and the Stark effect, as well as considerations bearing on the selection principle, have led to a knowledge of the values possessed by the *subordinate quantum number* k for the various terms, this quantum number determining the shape, and thus the eccentricity, of the orbital ellipse described by the valency electron. The values are :

$$(2) \quad \begin{cases} k = 1 \text{ for an } s\text{-term}, \\ k = 2 \text{ for a } p\text{-term}, \\ k = 3 \text{ for a } d\text{-term}, \\ k = 4 \text{ for an } f\text{-term}. \end{cases}$$

The terms for which k is to be put $= 5, 6, 7$ are known as g-, h-terms, and so on.

[1] The real discoverer of the "Bergmann series" was Saunders. The term "fundamental series" is, of course, very inappropriate.

[2] The f-terms are also often called b-terms (from the initial letter of the word Bergmann).

Observation shows that, the *lower* the *subordinate number of the initial state* on emission, the *sharper* is the spectral line produced. The lines in the second subordinate series are therefore the sharpest, those in the principal series being less sharp ; in the first (or diffuse) subordinate series the lines are, as the name indicates, ill defined, and in the Bergmann series they are least sharp of all.

Scheme (1) is a confirmation of the validity of the *selection principle*, discussed in an earlier section (§ 12), according to which only those spectroscopic combinations of terms are permissible where the *subordinate number changes by unity*. It is true that, in strong electrical fields, this principle does not hold, and lines then occur which appear to contradict it. For example, lines are observed in the generation of which the subordinate number remains unchanged ; frequently lines are seen which arise from a combination of two *p*-terms, or of two *s*-terms. In strong electrical fields combinations between *d*- and *s*-terms, and between *f*- and *p*-terms, may also occur, although the subordinate number then changes by two.[3]

It was found by *Rydberg* that the individual terms in the various series may be represented, in fair approximation, by the formula [4]

$$\nu = \frac{R}{(m+a)^2}, \tag{3}$$

where R denotes Rydberg's constant, and a is a constant which differs in value for the different succession of terms of an element. Furthermore, m denotes a variable whole number, its value commencing with 1 for the *s*-terms, 2 for the *p*-terms, 3 for the *d*-terms, and 4 for the *f*-terms.[5] The *variable* or *sequence number m*, however, must not be confused with the principal quantum number of the terms. It is only the most recent investigations of Bohr on the periodic system that have led to a knowledge of the principal quantum numbers of the terms, which we shall discuss in the following section.

The series relations appear most plainly with the elements of the first three groups, especially in that of the alkali metals.

[3] Series which are not in accord with the selection principle have been investigated chiefly by *Lenard* and *Stark*.

[4] The terms themselves are obtained on dividing eqn. (3) by c.

[5] Rydberg's formula was later improved by *Ritz*, the denominator being altered to

$$[m + a + \alpha f(1/m)],$$

where the function $f(1/m)$ is, in particular, either $1/m^2$ or $1/m$; α is a constant characteristic of all the terms of an element.

The *fundamental term* (*i.e.*, the numerically highest term, which corresponds to the lowest energy-level and measures the strength of the binding of the electron) is an *s*-term for the elements of the first and second groups. On the other hand, it has been shown that in the elements of the third group for which it has as yet been possible to arrange the lines in series, *viz.*, aluminium, gallium, indium, and thallium, the fundamental term is a *p*-term. This is confirmed by the fact that, in the absorption spectra given by the cold vapours of these metals (the atoms thus not being previously excited), only the two subordinate series occur, and the initial terms of the absorption lines in these are *p*-terms.

In the case of titanium and vanadium, which belong to the fourth and fifth groups respectively, it has been shown that the fundamental term is an *f*-term. A detailed analysis of the spectrum of *manganese* (seventh group) has been made by *Catalan*,[6] and the extremely complicated spectrum of *iron* (eighth group) has been attacked by *Walters* and by *Laporte*.[7] Of more than 2,000 lines in the iron spectrum, about 600 have been arranged in series: the fundamental term was found to be a *d*-term. Series relationships have also been detected in the spectra of the rare gases helium, neon, argon, and krypton; we shall discuss the helium spectrum more fully in a later section (§ 38). The *neon* spectrum has been analyzed by *Paschen*, who has succeeded in representing approximately 900 lines in it as combinations between four *s*-terms, ten *p*-terms, and twelve *d*-terms.[8]

Spark spectra of the first order have been investigated, apart from in helium, principally in the case of the *alkaline earths* (magnesium, calcium, strontium, and barium). The spark spectra of these metals do, in fact, exhibit a great similarity to the arc spectra of the corresponding alkali metals, the terms being quadrupled. The spark spectra of the alkali metals are just as rich in lines and as complicated as the arc spectra of the higher rare gases. With *aluminium Paschen* has discovered a *spark spectrum of the second order*,[9] and with *silicon Fowler* has obtained spark spectra of the

[6] *Cf. Phil. Trans. A.*, **223** (1922), p. 127.

[7] Walters, *J. Opt. Soc. Am.*, **8** (1924), p. 245; Laporte, *Zeitschr. f. Phys.*, **23** (1924), p. 135.—Recognition of the terms is rendered possible by the recurrence of exactly constant frequency differences, by regularities of interval, by regularities of intensity, and by observations on the Zeeman effect. It is worthy of note that, in the iron spectrum, the *s*- and the *p*-terms play a very small part in comparison with the *d*-, *f*-, *g*-, *h*-terms, *etc.*

[8] Paschen, *Ann. der Phys.*, **60** (1919), p. 405; and **63** (1920), p. 201.

[9] *Ann. der Phys.*, **71** (1923), p. 142.

second and of the *third orders*, as well as of the first order.[10] As is to be expected theoretically, the spectra of doubly ionized Al and of trebly ionized Si greatly resemble the arc spectrum of sodium, the terms being multiplied by nine and by sixteen respectively. Using the so-called explosive vacuum spark method, *Bowen* and *Millikan* have finally succeeded in obtaining the "*hot spark spectra*" of the "*stripped atoms*" for all the elements from Li to N, and from Na to Cl. Atoms are termed "stripped" when all the valency electrons except one have been removed; *e.g.*, in the case of chlorine, the stripped atom would give a spark spectrum of the sixth order. In addition to these "one electron" series, Bowen and Millikan have also obtained the hot spark spectra of the "two valency electron atoms" of B and C and of the elements from Al to Cl, as well as such spectra of the "three valency electron atoms" from Si to Cl, *i.e.*, of atoms which retained only two or only three of their valency electrons.[11]

Both in arc and in spark spectra, the majority of the terms appear as *multiplets*. In the arc spectra of the first and third groups, all terms except the invariably single s-terms are *doublets*. This applies also to the spark spectra of the the first order of the second group, and, in general, to the spectra of all "one-valency electron atoms." In the arc spectra of the second group there are *two systems of terms:* one system of *simple terms* and, in addition, a system in which all the terms except the single s-terms are *triplets*; the terms of the two systems occur in combination with one another. This applies also to the spectra of all "two-valency electron atoms." The rule appears to hold, quite generally, that with an *odd number* of valency electrons the terms are doublets, or quadruplets, sextets, or octets; whereas, with an *even number* of valency electrons, the terms are singlets and triplets, or quintets, septets, or nonets.

In doublets and triplets the individual terms are denoted by the addition of a *suffix*, which, in accordance with a notation generally adopted since about 1925 (and opposed to the one previously in use), is identical with the *inner quantum number* introduced by *Sommerfeld*. Thus the higher energy-level is associated with the higher suffix. The system of simple terms occurring with an even number of electrons is distinguished by capital italic letters (S, P, D, F). The previously mentioned sequence number of the terms (which

[10] *Proc. Roy. Soc., A* **103** (1923), p. 413.

[11] *Vide* Bowen and Millikan, Phys. Rev., 1923 *ff., passim.*

commences with the s-terms[12] at 1, with the p-terms at 2, with the d-terms at 3, and with the f-terms at 4) is customarily written before the capital or small letters. Thus we speak of a $3P$-, or $2s$-, or $5p_2$-, or $4d_3$-term, *etc.*

When two terms belonging to a doublet or a triplet are combined with any third term, the *stronger line* always corresponds to the one of the two terms with the *larger suffix*. *E.g.*, the line $2p_2 \rightarrow 1s$ is always stronger than the line $2p_1 \rightarrow 1s$. The *frequency difference* of multiple terms invariably decreases with increasing sequence number, and converges to zero. For the same sequence number, the frequency difference of the p-doublet is always greater than that of the d-doublet, and this, in its turn, is greater than that of the f-doublet; the same holds good for triplets.

Within one and the same group of the periodic system, higher periods are always associated with larger frequency differences. With low atomic numbers, however, the terms do not yet appear multiple. Thus, with lithium and sodium the d-term is still simple, with magnesium it is not yet triple, and it is not until barium is reached that the f-term appears as a triplet.

The formation of spectroscopic combinations between the individual terms is governed by a *selection rule* formulated by *Sommerfeld* and *Landé*. This rule is identical with the principle expressed by § 31 eqn. (1), with the further condition that the transition $0 \rightarrow 0$ is excluded. The *simple terms* S, P, D, F are allotted the inner quantum numbers 0, 1, 2, 3. The assignment of j-values to *doublet terms* follows a *parallel* usage to the case of the *Röntgen terms*, according to the following scheme :—

$$(4) \qquad \begin{cases} K \quad L_{11}\ L_{21}\ L_{22} \quad M_{11}\ M_{21}\ M_{22}\ M_{32}\ M_{33} \\ 1s \quad 2s\ \ 2p_1\ 2p_2 \quad 3s\ \ 3p_1\ 3p_2\ 3d_2\ 3d_3 \\ \qquad N_{11}\ N_{21}\ N_{22}\ N_{32}\ N_{33}\ N_{43}\ N_{44} \\ \qquad 4s\ \ 4p_1\ 4p_2\ 4d_2\ 4d_3\ 4f_3\ 4f_4. \end{cases}$$

Inner quantum numbers are allotted to triplet terms in an analogous manner.

By way of illustrating this rule, let us consider the principal series and the two subordinate series in the *arc spectra of the alkali metals*. According to Sommerfeld's selection

[12] The sequence numbers of the s-terms are frequently given the values 1·5, 2·5, 3·5 . . . instead of 1, 2, 3 . . . because the constant in Rydberg's equation (3) is approximately equal to 1/2 for the s-terms.

principle, the lines in the *principal series* must be *doublets*, for the s-term is simple and, because its inner quantum number is 1, transitions from both the p_1- and the p_2-levels are possible. Since the p_2-term corresponds to the higher level, the line $p_2 \rightarrow s$ has the larger frequency and also, in view of what has been said above, the greater intensity. Moreover, with increasing sequence number, the frequency difference of the p-doublet will, as we have seen, become steadily smaller. The lines in the principal series are therefore composed of constantly narrowing doublets, in which the line with the *smaller wave-length* has the *greater intensity*. The series limit, which is represented by the simple s-term, is the same for both components of the doublet.

In the second or *sharp subordinate series*, which, in emission spectra, is due to transitions from a simple s-level to a double p-level, the lines are obviously again doublets. Since, however, the p-term now represents the final level (instead of the initial level as before), the smaller frequency corresponds to the line belonging to the p_2-term. Hence, in contrast to the principal series, in the doublets of the sharp subordinate series the *line of greater wave-length* is the *stronger*. Since the constant term for the sharp subordinate series is double and the variable term simple, the *frequency difference of the doublets* is *constant* for the whole series, so that the series actually has two series limits.

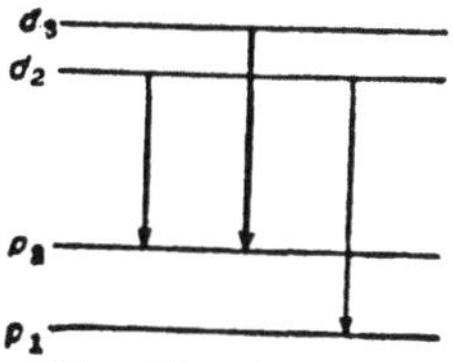

FIG. 43.—Optical triplet.

In the lines of the *diffuse subordinate series*, Sommerfeld's selection principle shows that *three* transitions are always possible, *viz.*, $d_3 \rightarrow p_2$, $d_2 \rightarrow p_2$, and $d_2 \rightarrow p_1$. According to the rule given, the lines of greatest intensity must correspond to the transition $d_3 \rightarrow p_2$, for the suffixes are then the largest. This transition is associated with a larger frequency than the transition $d_2 \rightarrow p_2$, but with a smaller frequency than the transition $d_2 \rightarrow p_1$, because, as we have seen, the frequency difference of the d-doublet term is always less than that of the p-doublet term (*cf.* Fig. 43, where the different intensities of the lines are also symbolically represented). The lines in the diffuse subordinate series are therefore *triplets*, in which the central line exhibits the greatest intensity.[13]

[13] As is immediately obvious, however, the series limit is only double; the frequency difference is that of the p-doublet.

In every spectrum there is, of course, not merely one single principal series, a single diffuse and a single sharp subordinate series, but there is a multiplicity of such series; for example, the final level of a principal emission series may be any arbitrary *s*-term, and thus not only 1*s*, but also 2*s* or 3*s*, and so on. By principal and subordinate series in the narrower senses of the words, however, we always understand the series whose final levels on emission are the lowest *s*-level (1*s*) and the lowest *p*-doublet level (2*p*) respectively.

TABLE XI

The First Lines in the Principal Series of the Alkali Metals
(Wave-lengths in Å.U.)

Li	Na	K	Rb	Cs
6708·2	D_1 5895·930	7698·97	7947·6	8943·6
—	D_2 5889·963	7664·94	7800·2	8521·1

The first and *strongest line* of *longest wave-length* in the *principal series* (in the narrower sense of the word) lies in the *visible* region of the spectrum for all the alkali metals except cæsium. The wave-lengths are given in Table XI, from which it may also be seen that the frequency difference of the doublet increases with increasing atomic number. (With lithium the *p*-term is not yet double, and hence the line is simple.) For sodium the first line of the principal series lies in the yellow; it is the well-known Fraunhofer *D*-line, the component (D_2) of shorter wave-length being stronger (about twice as strong) than the other component (D_1), in accordance with the previous statements. For Li, K, and Rb the first line lies in the red, whereas for Cs it is situated in the infra-red close to the border of the visible region.[14]

Since the first line of the principal series is determined by the difference of the terms 1*s* and 2*p*, the frequency of this line is equal to the difference between the limiting frequency of the principal series and the common limiting frequency of the two subordinate series. This rule is known as *Rydberg and*

[14] By way of comparison, it may be noted that the first line in the "principal series" of hydrogen lies far in the ultra-violet, at 1,216 Å.U.

Schuster's rule, these two investigators having established it long before the birth of Bohr's theory.

In the elements of the first group there are no further levels between the $2p_1$-level and the $1s$-level, which represents the normal state, and therefore the frequency of the line

TABLE XII

The Principal Series of Sodium

($mp \rightarrow 1s$; wave-lengths in Å.U.)

$m = 2$ 5895·930 D_1 5889·963 D_2	3 3302·94 3302·64	4 2853·038 2852·821	5 2680·443 2680·335	6 2593·927 2593·828	7 2543·875 2543·817
8 2512·210 2512·128	9 2490·733 —	10 2475·533 —	11 2464·397 —	12 2455·915 —	13 2449·393 —
14 2444·195	15 2440·046	16 2436·627	17 2433·824	18 2431·433	19 2429·428
20 2427·705	21 2426·217	22 2424·937	23 2423·838	24 2422·856	25 2421·997
26 2421·233	27 2420·520	28 2419·922	29 2419·380	30 2418·893	31 2418·454
32 2418·062	33 2417·695	34 2417·362	35 2417·058	36 2416·779	37 2416·518
38 2416·271	39 2416·046	40 2415·838	41 2415·651	42 2415·474	43 2415·305
44 2415·147	45 2415·006	46 2414·872	47 2414·746	48 2414·627	49 2414·518
50 2414·411	51 2414·313	52 2414·218	53 2414·131	54 2414·050	55 2413·971
56 2413·910	57 2413·873	58 2413·837	— —	— —	— —

$2p_1 \rightarrow 1s$, multiplied by the elementary quantum of action, gives the *excitation energy* of the atom. In accordance with the selection rule, it is a property of the $2p_2$-level as well as of the $2p_1$-level that, on the emission of light, *only a transition from that level back to the normal state is possible.* An atom which *absorbs* light of one of these two wave-lengths, and thereby becomes excited, can thus *emit* light only of the *same* wave-length on its return to the normal state. The double line $2p \rightarrow 1s$ is therefore also termed the *resonance line* of the alkali metals.

The *second line* in the principal series (*i.e.*, the doublet $3p \rightarrow 1s$) is situated in the ultra-violet in the cases of Li and Na ; for K it is the well-known violet doublet ; for Rb it is likewise a doublet in the violet, but of greater wave-length ; and, finally, for Cs it is the well-known blue doublet, from which cæsium takes its name.[15] This second line also

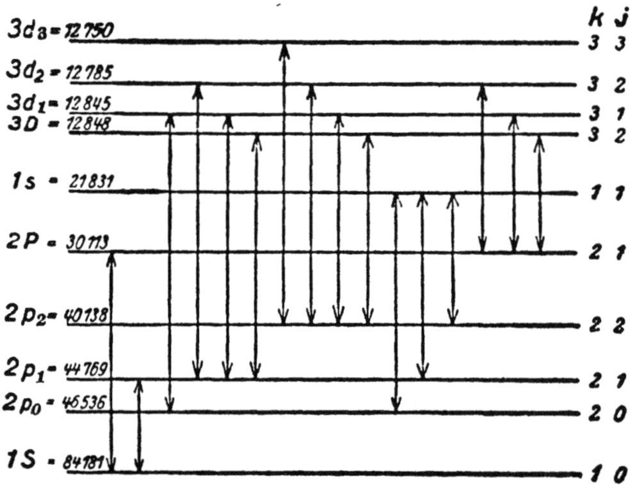

FIG. 44.—Energy-levels for mercury.

shows clearly the increase in width of the line doublet with the atomic number.

The *principal series in the arc spectrum of sodium*, in which it has been possible to measure no less than 57 lines, is given as an example of an optical series in Table XII.

In the case of the *elements of the second group*, the *arc spectra* are naturally considerably more complicated than those of the alkali metals. For, quite apart from the occurrence of triplet terms, we find in these spectra two different systems of terms between which spectroscopic transitions occur. The fundamental term, *i.e.*, the numerically highest one, belongs to the system of simple terms ; it is the $1S$-term, with the subordinate number which determines the eccen-

[15] *Cæsius* is the Latin word for blue.

tricity of the orbit equal to unity. In the emission spectrum it forms the final level for two principal series, one of which arises in transitions from P-levels, and the other in transitions from p_1-levels. (Transitions from a p_0- or p_2-level are excluded by the selection rule previously enunciated.) Both the $2P$-level and the $2p_1$-level are found to be lower than the $1s$-level. Hence, on the emission of monochromatic light, only the return to the fundamental state $1S$ is possible from the $2P$- and the $2p_1$-level. To these two transitions correspond the *two resonance lines* in the arc spectra of the elements of the second group (*viz.*, $2P \rightarrow 1S$ and $2p_2 \rightarrow 1S$).

In Fig. 44 we see represented the most important terms and spectral lines in the arc spectrum of *mercury*, which belongs to the second group. The terms corresponding to the various levels are shown, so that the wave-length of a spectral line (in centimetres) is simply equal to the reciprocal of the difference between the terms of the levels between which the transition occurs. For example, the wave-lengths of the two resonance lines of mercury are the reciprocals of (84,181—44,769) and (84,181—30,113), or 2,537 and 1,849 Å.U.; both lines therefore lie in the ultra-violet. Fig. 44 also clearly shows the validity of Sommerfeld and Landé's selection rule; the values of the subordinate quantum number (k), which determines the eccentricity of the orbit, and of the inner quantum number (j) are given on the right-hand side of the figure.

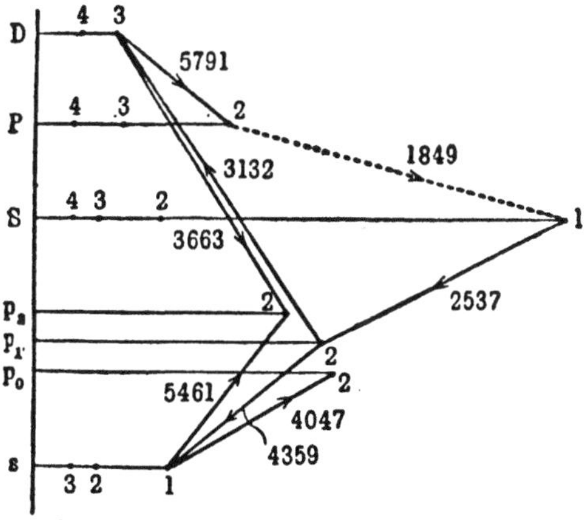

FIG. 45.—Arc spectrum of mercury.

Another method of representing the arc spectrum of mercury is that shown in Fig. 45. The numerical values of the terms are plotted to scale along horizontal lines drawn from the vertical zero-line (the numerals 1, 2, 3, 4 denote the sequence numbers). The connecting lines represent the transitions permitted by the selection rule, and the numbers written along these lines give the wave-lengths in Å.U.

(The signification of the arrows will be discussed in a later section.)

It has been theoretically deduced that *simple integral ratios* should exist between the *intensities* of the individual lines in a spectral multiplet, and this deduction has been experimentally confirmed by *Ornstein*. For example, the intensity ratio for the lines of a doublet is 2 : 1, whilst that for the lines of a triplet is 5 : 3 : 1; and analogous relations hold for the lines of other multiplets.[16]

[16] For further details *vide* Sommerfeld's "*Atomic Structure and Spectral Lines*," 4th edn., ch. viii., § 5.

Translator's Note on the Classification of Optical Spectra and the Notation of Spectral Terms.

In the method of classifying optical spectral terms described in this and the following sections, it is assumed that the atom can be divided into a single photo-electron and a core, and the arrangement of the terms in series is based on the value of the subordinate quantum number k associated with this electron. This mode of classification is consequently known as the "k-classification."

There is, however, a second method of classifying terms which has recently come into general use. It has been pointed out by Heisenberg * that the k-method applies only when the interaction between the photo-electron and the other external electrons is small as compared with that between the photo-electron and the core. With the more complex atoms and spectra, it is not possible to select any one electron as the photo-electron, and we must rather regard the atom as being divided into a shell of external electrons and a rare-gas core. In these circumstances, the interaction between the external electrons will be large as compared with the interaction between them and the core, and we must adopt a classification based on the value of the quantum number l associated with the resultant angular momentum of the shell of external electrons : this is known as the "l-classification."

The k-classification holds for the relationships between the optical series and between them and the periodic system, whereas the l-classification applies to the investigation of multiplets, the Zeeman effect, and selection principles. Only in the case of the simplest spectra do both methods lead to the same result. Inasmuch as the more complex spectra have been resolved almost exclusively by the l-method (the spectrum of Ne is an exception), it seems logical to prefer the l-classification in general, and we shall therefore briefly explain the notation used in connection with it, which has been proposed by Russell and Saunders, and developed by Sommerfeld.

The letters $S, P, D, F, \ldots$ are used to denote the values 0, 1, 2, 3, . . . of l. The general symbol

$$^{r}X_{j}$$

is written for a term, where X represents one of the letters S, P, D, etc., r is the multiplicity of the term, and j is the inner quantum number. When the subordinate quantum number k of the photo-electron is definitely determined, the symbol n_k may be prefixed (n representing, as usual, the principal quantum number). The complete characterisation of a term would demand the symbolical representation of the orbit of every electron in the atom. This is

* W. Heisenberg, *Zeits. f. Phys.*, 32 (1925), p. 841.

§ 34. The Grouping of the Electrons in the Atoms.

Closely related to the theory of spectra is the problem of the *grouping* of the electrons around the atomic nucleus in the atoms of the different elements. The first successful attempt at such an *interpretation of the periodic system of the elements on the basis of the electron theory* was made by *Bohr* in 1922. Bohr's views have been further developed principally by *Stoner* (1924), but a complete solution of this problem of atomic theory has not yet been reached.

According to Bohr, the electrons surrounding the nucleus may be divided into *different quantum groups*, and he establishes a certain relationship between these groups and the periodic system by assuming that, in the normal state of the atoms, two-quantum orbits first appear in the second period, three-quantum orbits in the third period, and so on.

The hypothesis recently advanced by Stoner states that each group may be further subdivided into *as many sub-groups* as there are *Röntgen levels* in the Röntgen spectrum of the atom concerned, the *number of electrons* in each of these sub-groups, when fully completed, being equal to *twice the inner quantum number* of the corresponding Röntgen level.

Thus, the total number of electrons in a completed n-quantum group is given by

$$(1)\quad \begin{cases} Z = 2\,[1 + (1 + 2) + (2 + 3) + (3 + 4) + \dots \\ \qquad\qquad + \{(n - 1) + n\}]. \end{cases}$$

If, following the usual notation of the Röntgen levels, we call the one-quantum group the K-group, the two-quantum group the L-group, *etc.*, we find the *number of electrons* in the *completed* K-, L-, M-, and N-groups to be 2, 8, 18, and 32, and it is obvious that these numbers are intimately

scarcely possible, but, following Russell and Saunders, we may indicate the general species of the core by adding dashes to the symbol: thus we speak of undashed, dashed, and twice-dashed terms.

In the l-classification, transitions are governed by the selection principles

$$\Delta l = \pm 1 \;(0) \text{ and } \Delta j = \pm 1 \;(0).$$

For details as to the grouping of electrons, with a summary of our present knowledge, the reader may be referred to a recent paper by Grimm and Sommerfeld.†

† H. G. Grimm and A. Sommerfeld, *Zeits. f. Phys.*, **36** (1926), p. 36.

TABLE XIII

Grouping of the Electrons in the Atoms of the Rare Gases

Sub-group	K	L_{11}	L_{21}	L_{22}	M_{11}	M_{21}	M_{22}	M_{32}	M_{33}	N_{11}	N_{21}	N_{22}	N_{32}	N_{33}	N_{43}	N_{44}	O_{11}	O_{21}	O_{22}	O_{32}	O_{33}	P_{11}	P_{21}	P_{22}
Number of electrons when completed	2	2	2	4	2	2	4	4	6	2	2	4	4	6	6	8	2	2	4	4	6	2	2	4
2 Helium .	2																							
10 Neon .	2	2	2	4																				
18 Argon .	2	2	2	4	2	2	4																	
36 Krypton .	2	2	2	4	2	2	4	4	6	2	2	4												
54 Xenon .	2	2	2	4	2	2	4	4	6	2	2	4	4	6			2	2	4					
86 Emanation .	2	2	2	4	2	2	4	4	6	2	2	4	4	6	6	8	2	2	4	4	6	2	2	4

connected with those of the elements in the six periods of the periodic system, *viz.*, 2, 8, 8, 18, 18, and 32.[1]

Now, as was mentioned in § 31, investigation of the *Röntgen spectra* has shown that the *emanation atom* evidently possesses 1 *K*-, 3 *L*-, 5 *M*-, 7 *N*-, 5 *O*-, and 3 *P*-levels, so that, in this atom, the one-, two-, three-, and four-quantum groups are completely developed, but the five-, and six-quantum groups only partially so.

Similarly, a consideration of the Röntgen spectra of the elements immediately following xenon (*cf.* the earlier Fig. 40) leads to the assumption of the existence of 1 *K*-, 3 *L*-, 5 *M*-, 5 *N*-, and 3 *O*- levels in the *xenon atom*, whilst the *krypton atom* apparently possesses 1 *K*-, 3 *L*-, 5 *M*-, and 3 *N*-levels. In this way we obtain the scheme of *electron grouping in the atoms of the rare gases*, which is reproduced in Table XIII ; a short summary is given in Table XIV.

TABLE XIV

Number of Electrons in the Different Quantum Groups in the Atoms of the Rare Gases

Rare gas	Principal quantum number of group						Total number of electrons
	1	2	3	4	5	6	
Helium . .	2						2
Neon . . .	2	8					10
Argon . .	2	8	8				18
Krypton . .	2	8	18	8			36
Xenon . .	2	8	18	18	8		54
Emanation . .	2	8	18	32	18	8	86

A special position is occupied, not only by the rare gases, but also, to a certain extent, by those elements in which the *development of a sub-group is completed.* Such should be the second and the fourth elements in each period, thus on the one hand, the elements in column II A of the periodic system (Be, Mg, Ca, Sr, Ba, Ra), and, on the other hand, the elements C and Si. All these elements are distinguished by the property of *diamagnetism* (*cf.* § 39), and also by the

[1] The complete five-quantum *O*-group would contain 50, the complete ix-quantum *P*-group 72 electrons, and so on.

possession of optical *fundamental terms* for which the *inner quantum number* $j = 0$. For example, the fundamental term for the atoms in column II A is an S-term, whilst, for C and Si, it is probably a p_0-term.

Again, the atoms of the elements immediately *succeeding* those in which a sub-group is completed have a fundamental term such that the *quantum numbers* k and j are identical with those of the *new sub-group about to be developed.* Thus the fundamental term of the alkali metals is an s-term, and that of the elements in the third column a p_1-term ($k = 2$, $j = 1$), as is also the fundamental term for ionized carbon.

Since, as we saw in § 33, the fundamental term of the alkali metals is always associated with the subordinate number $k = 1$, the valency electron in these elements evidently describes about the nucleus as focus an orbit which, among all conceivable orbits with the same principal quantum number, exhibits the *greatest eccentricity* and the *most elongated shape* (*cf.* the earlier Fig. 9 in § 12).

On the other hand, however, it is evident from Table XIV that, in lithium, the valency electron cannot possibly be situated in the completed one-quantum group of electrons, and in sodium there is no vacancy for it in the completed two-quantum group. On account of the great similarity between the alkali metals, we must hence quite generally conclude that, in all these metals, the valency electron describes a path whose *principal quantum number* is *larger by one* than the highest principal quantum number normally occurring in the preceding rare gas. Thus, in the normal state, the valency electron would describe in Li a two-, in Na a three-, in K a four-, in Rb a five-, and in Cs a six-quantum orbit.

Although these orbits of the valency electrons possess quite different principal quantum numbers, it follows nevertheless from the optical spectra that the *energy binding* the valency electron to the nucleus has *practically the same value* for all the alkali metals. This may be seen in Table XV, which also shows that the binding in all the alkali metals is weaker than the binding of the electron in the normal one-quantum hydrogen atom, but stronger than that in the excited two-quantum hydrogen atom. The binding energy is represented in Table XV by the *fundamental term* belonging to the principal series, for this, multiplied by the product hc, gives that energy.[2]

[2] The hydrogen terms given in Table XV are obtained on dividing the Rydberg term by the square of the principal quantum number.

TABLE XV

Fundamental Terms of the Alkali Metals

Element	Fundamental term	Principal quantum number of the fundamental term	Hydrogen term with the same principal quantum number
—	—	1	109,678
3 Li	43,484	2	27,419
11 Na	41,445	3	12,186
19 K	35,006	4	6,855
37 Rb	33,685	5	4,387
55 Cs	31,407	6	3,047

The nearly identical value of the binding energy in the different alkali metals is due to the *large eccentricity* of the orbits of the *valency electrons*. The very elongated orbit of the valency electron lies, indeed, for the most part far outside the electron groups with smaller principal quantum numbers: thus, most of the orbit is located in regions from which the nuclear charge is almost completely screened off, except, as we have seen, for a single elementary quantum. At *perihelion*, however, owing to the great eccentricity of the orbit, the valency electron approaches *extremely close to the nucleus*, to within distances which are at any rate smaller than the radius of the 2_2-orbit, and most probably smaller even than the radius of the 1_1-orbit.[3]

The valency electron, therefore, penetrates during its revolution very deeply into the interior of the atom, and thus enters regions in which there is little or no screening of the nuclear charge, so that the full or almost full strength of the nucleus acts upon it. Hence, although, in the cæsium atom, for example, the valency electron normally describes a six-quantum orbit and consequently travels very far from the nucleus, and although, during the greater part of its revolution, only a single elementary quantum of the total nuclear charge acts upon it, yet, on the other hand, along a very small portion of its orbit the valency electron is attracted at extremely close range by a charge of 55 or 53 elementary quanta. The constant energy [4] along the orbit is therefore,

[3] *Cf.* the earlier Fig. 9 in § 12.

[4] In accordance with the conservation principle, this energy is, of course, constant only inasmuch as the mutual action between the valency electron and the other electrons may be neglected.

in absolute amount, much larger than in a six-quantum hydrogen atom, although most of the orbit is that of a "hydrogen-analogue," inasmuch as only a single elementary quantum of the nuclear charge is effective.

Besides the rare gases and the alkali metals, a special position is also occupied in the periodic system by the three elements which form group I B. These are *copper*, *silver*, and *gold*, which, both chemically and spectroscopically, exhibit far-reaching similarities to the alkali metals. The atomic numbers of these three metals are lower by eight than those of the alkali metals which commence the following periods.[5] This fact suggests that the grouping of the electrons in these three metals may be the same as in the next higher rare gases, except that, in each case, the highest-quantum group, which in the atoms of each of the rare gases contains eight electrons, will be represented in copper, silver, and gold by only a *single electron*. In the last eight places in each of the periods, the highest-quantum group would then be successively completed from one to eight electrons. The description of a very eccentric orbit of subordinate quantum number one by the valency electron in copper, silver, and gold follows both from their similarity to the alkali metals and from the fact of experimental spectroscopy that the optical fundamental term for copper, silver, and gold is an *s*-term.

§ 35. The Completion of the Inner Electron Groups.

As is clearly shown by the earlier Table XIV, an essential assumption of the modern theory of chemical periodicity is that the *development of the various quantum groups of electrons* from the principal quantum number three onwards takes place *in stages* and covers *several periods*. Thus, in the fourth period, the building up of the outermost four-quantum group of electrons must appear *interrupted* by the *completion* of the *inner* three-quantum group from 8 to 18 electrons. Similarly, in the fifth period, the development of the outermost five-quantum group is interrupted by the growth of the four-quantum group from 8 to 18 electrons, and in the sixth period the completion of the fifth group from 8 to 18 electrons interrupts the development of the outermost six-quantum group. Moreover, a further apparent delay in the development of the peripheral group in the sixth period must be caused by the completion, still more deeply within the atom, of the four-quantum group from 18 to 32 electrons.

[5] Including the unknown element 87 with the alkali metals.

From the point of view of the *quantum theory*, a clue to this *development of the inner groups of* electrons may most readily be found from *optical spectra*. Let us consider three atoms, each of which possesses 18 electrons in the configuration of the argon atom and, in addition, a nineteenth, more loosely bound electron ; and let us suppose the nuclear charges of the atoms to be 19, 20, and 21 elementary quanta. These atoms would therefore be a *neutral potassium atom*, a *singly ionized atom of calcium* (Ca^+), and a *doubly ionized scandium atom* (Sc^{++}). In each of the three cases we shall suppose the loosely bound nineteenth electron to occupy one of *three different types of orbit, viz.*, a 4_1-, a 4_2-, and a 3_3-orbit ; and we then ask ourselves how strong the *binding* of the electron will be.

As regards the neutral potassium atom, the *arc spectrum of potassium* gives us some help. We find that the highest s-term (corresponding to the lowest s-level) is 35,006, the highest p-term (a doublet) 22,021 and 21,963 respectively, and the highest d-term 13,471 and 13,468 respectively. From the considerations of the previous section (§ 34), we know that the highest s-term, as fundamental term, corresponds to a 4_1-orbit. We must likewise assume that the highest p-term (for which the subordinate number is equal to two) is associated with the principal quantum number four, seeing that the 3_2-sub-group was completed with argon, so that there is no place left in that sub-group for another electron. On the other hand, we may well ascribe the principal quantum number three to the highest d-term in potassium (the subordinate number being equal to three), for, in the argon atom, the 3_3-sub-group is quite unoccupied. This conception is also justified by the approximate agreement of the highest d-term for potassium with the three-quantum hydrogen term,[1] as might be expected from the circular form of the 3_3-orbit. In Fig. 46 the values of the three terms for potassium are shown, calculated, however, in volts according to the relation[2]

$$(1)\qquad \lambda V = 1{\cdot}234 \times 10^4 \,(\lambda \text{ in Å.U., } V \text{ in volts}).$$

The nature of the bindings in the singly ionized calcium atom is revealed by the *spark spectrum of calcium*, which is analogous to the arc spectrum of potassium with fourfold terms. Dividing the terms by four, we find for the highest

[1] *Cf.* the last column in Table XV.
[2] § 20 eqn. (3).

s-term in the spark spectrum of calcium (*i.e.*, the 4_1-term) the value 23,930, for the highest *p*-term (*i.e.*, the 4_2-term) 17,632 and 17,576 respectively, and for the highest *d*-term (*i.e.*, the 3_3-term) 20,517 and 20,502 respectively. In Fig. 46 are represented the three terms for singly ionized calcium (Ca^+) calculated in volts.

Now Fig. 46 clearly indicates a considerable difference between the two spectra. In the arc spectrum of potassium, the 3_3-binding is weaker not only than the 4_1-, but also than the 4_2-binding. In the spark spectrum of calcium, however, the 3_3-binding is *stronger* than the 4_2-binding. Whereas the 4_1- and the 4_2-terms for Ca^+ are considerably *smaller* than in K, the 3_3-term in Ca^+ is considerably larger than in K. We do not, it is true, as yet know the spectrum of *doubly ionized scandium*. With the help of a very arbitrary *extrapolation*, however, as shown in Fig. 46, we may suppose that in doubly ionized scandium the 3_3-term is *larger* than the 4_1- and the 4_2-term.

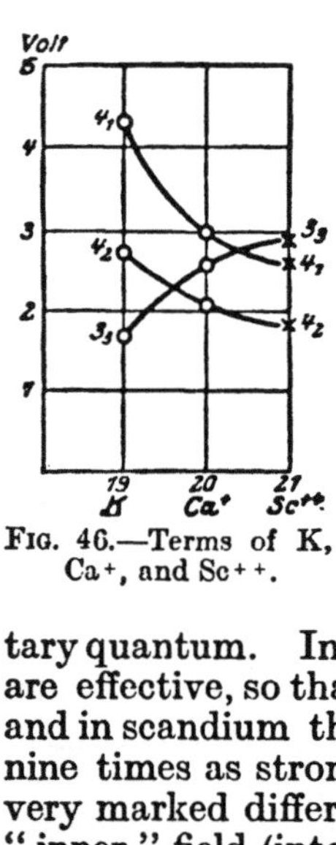

FIG. 46.—Terms of K, Ca^+, and Sc^{++}.

This rapid growth of the 3_3-term becomes intelligible to us when we remember that, *outside* the 18 electrons which are grouped as in the argon atom, the atomic nucleus of potassium has approximately only one effective elementary quantum. In calcium, however, two elementary quanta are effective, so that the binding energy is four times as large, and in scandium three elementary quanta, the binding being nine times as strong. Whereas, therefore, with potassium a very marked difference exists between the " outer " and the " inner " field (into which the eccentric orbits of subordinate numbers one and two penetrate [3]), this difference is not so large with calcium, and is still smaller with scandium. Hence, although with potassium the 4_1-term is about twice as large as the 3_3-term, the binding for scandium of a nineteenth electron in a 3_3-orbit is stronger than in an eccentric 4_1-orbit, which, although it penetrates in part into the interior of the atom, yet carries the electron much further from the nucleus than is the case for a circular 3_3-orbit.

Again, the grouping of the electrons in a neutral atom must evidently be exactly the same as if all the electrons

[3] *Cf.* § 34.

surrounding the nucleus were torn away, and the atom were *reformed* by *successive binding of electrons to the nucleus.* In this process each electron would, in a sense, "fall as low as possible," *i.e.*, it would occupy the lowest free energy-level. We may therefore suppose that, in such a successive building up of a scandium atom, the nineteenth electron would take up a 3_3-orbit, and thus that scandium is the first element in which a 3_3-orbit is normally present. The *completion of the three-quantum group* from 8 to 18 electrons therefore evidently commences with *scandium* and finishes with *copper.* The process of completion thus extends over *nine places* (Nos. 21–29); the *increase of ten electrons* in the inner electron group is explained by the fact that the outermost, four-quantum, electron group is represented in calcium by two electrons, but in copper by only one.

Since the 18 elements in the *fifth period* are complete homologues, in the same order, of the 18 elements in the fourth period, we must assume that in the fifth period the completion of the inner (*i.e.*, four-quantum) group of electrons also commences with the third element, *yttrium,* of atomic number 39, and ends with *silver*, of atomic number 47.

In the *sixth period*, the completion of the five-quantum group from 8 to 18 electrons evidently commences with *lanthanum* which, as third element in this period, is an exact homologue of scandium and yttrium, and ends with *gold.* In this case the completion extends over 23 instead of 9 places, *viz.*, from No. 57 to No. 79. Presumably, therefore, the four-quantum group also is completed between lanthanum and gold from 18 to 32 electrons.

In the elements of the sixth period, however, this four-quantum group lies fairly *deep within the atom.* Hence, during its completion, the *chemical* and also those *physical properties* such as valency and optical-spectral behaviour, which are conditioned by the *peripheral electrons*, must remain practically *unchanged.* In point of fact, the elements following after lanthanum, the so-called *rare earths*, closely resemble one another and lanthanum in chemical and in important physical properties, whereas lanthanum is very dissimilar from the immediately preceding element, barium. We must thus assume that the completion of the four-quantum group from 18 to 32 electrons *begins* with the element next after lanthanum, *cerium* (58). Corresponding to the difference between the numbers 32 and 18, there must naturally be a total of 14 elements resembling lanthanum, or, including this element, *fifteen rare earths* altogether.

The agreement of the peripheral atomic properties does, in fact, extend over 15 successive elements,[4] beginning with lanthanum and ending with element 71, *lutecium*. At the time when Bohr established his theory of the periodic system, element 72 was not yet known. Up till then it had commonly been supposed that it must also be a rare earth, and searches were therefore made for it in those minerals in which other rare earths had been found. It followed from the new theory of Bohr, however, that the unknown element could not possibly be a rare earth (because the theory does not permit of more than 15 rare earths), but rather that it must be a *homologue of zirconium*, which occupies the fourth place in the fifth period. Since homologous elements frequently occur together in nature, the idea suggested itself of looking for the unknown element in zirconium minerals. It constituted a brilliant triumph for Bohr's new theory when, in 1922, *Hevesy* and *Coster* actually succeeded in discovering the new element in zirconium minerals; in honour of the city of Copenhagen, where the discovery was made, they named it "*hafnium*." In point of fact, hafnium is by no means rare; there is, for example, only about five times more nickel than hafnium in the earth's crust. It can only be ascribed to its close resemblance to zirconium that hafnium was not discovered long ago, and also that it has not yet been possible to prepare it in a pure state.[5]

In the *seventh period* of the system of the elements, the six-quantum group must be completed from 8 to 18, and, in addition, the five-quantum group from 18 to 32 electrons. From the homology of actinium and lanthanum, of protoactinium and tantalum, and of uranium and tungsten, we must conclude that the completion of the second outermost group of electrons (in this case, the six-quantum group) also begins in the seventh period with the third element, actinium, just as in the fourth, fifth, and sixth periods. On the other hand, up to and including uranium, the number of electrons in the five-quantum group further within the atom does not increase.

Fig. 47 represents the periodic system of the elements from the point of view of Bohr's new theory. The seven periods are arranged in vertical columns. Series of elements within which an inner group of electrons is understood to be completing itself are enclosed in rectangles; in the sixth

[4] Including, it is true, the element 61, which is not yet known.

[5] The discovery was made by Röntgen spectroscopy. *Vide Nature*, 111 (1923), *passim*.

period there are two such rectangles, one within the other.[6] Homology of elements in different periods is indicated by tie-lines, but lines are not drawn between elements which, despite the same valency, occupy different positions with respect to the rectangles ; thus, there is none between Al and Sc, though there is between Sc and Y.

Elements with incomplete inner atomic structure exhibit various peculiarities lacking in the other elements. They

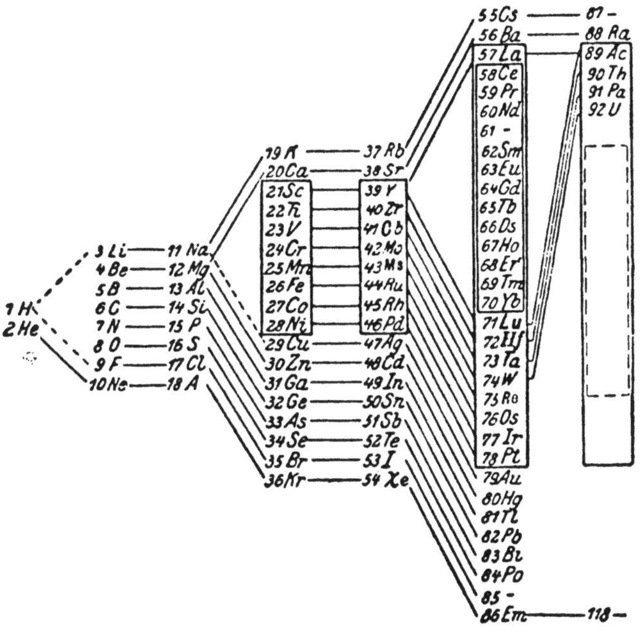

FIG. 47.—Periodic system, according to Bohr.

are frequently *multivalent ;* they form *coloured salts ;* and, as a rule, they are *paramagnetic.*

The multivalency is evidently due to the possibility that electrons belonging to an incompletely developed inner group may be just as *loosely* bound as the peripheral valency electrons.[7] Hence, like the latter, they are also able to absorb

[6] The rectangles in Fig. 47 end before the elements in which the inner groups of electrons are completed. In the seventh period the inner rectangle is obviously quite arbitrarily drawn ; we know only that the elements up to uranium inclusive lie outside it.

[7] We have, moreover, to assume such a loose binding of the inner electrons in the elements which, though lying outside the rectangles in Fig. 47, yet

quanta of *visible* light, so that their ions are coloured, despite the absence of valency electrons. Finally, as we shall discuss more fully in a subsequent section (§ 39), paramagnetism is ascribed to an incomplete balance in the magnetic moments of the electronic orbits in the individual atoms, and thus likewise appears as a consequence of asymmetry in the internal structure of the atom.[5]

A direct *experimental confirmation* of Bohr's attempts to explain chemical periodicity and homology from the point of view of the quantum theory is afforded by the *curves* representing the *dependency of the square roots of the Röntgen-spectroscopic terms of the elements on the atomic numbers* (*cf.* Fig. 42 in § 31).

In accordance with Moseley's Law, the form of these curves should be exactly *rectilinear*, provided that an increase in the atomic number by unity is accompanied by a like increase in the *effective fraction of the nuclear charge* for the term concerned, and a corresponding increase in the *binding energy*. During the development of an inner electron group, however, the increase in the binding energy and hence of the term must at any rate be interrupted, or at the very least, retarded. This is due to the fact that the increase in the nuclear charge is entirely or partially counterbalanced by a similar increase of one in the number of *inner electrons screening* the nucleus. Hence, *during the development of an inner group of electrons* of given quantum number, the *curves* for the terms of higher quantum number should run *horizontally*, or at least at a *smaller angle of inclination* to the axis of abscissæ.

That such is the case is clearly shown by the earlier Fig. 42 and by Fig. 48, in which part of Fig. 42 is reproduced on a larger scale. Vertical lines underneath both figures indicate the elements in which new types of orbit first occur, whilst the development of an inner group of electrons is shown by a horizontal line. In both figures the horizontal course of the five-quantum *O*-terms during the completion of the four-quantum group from 18 to 32 electrons is clearly visible. Moreover, during this completion, the curves for the *N*-terms run less steeply, for the addition of new electrons

border on them. Thus *copper* is both *univalent* in *cuprous* compounds and *divalent* in *cupric* compounds. Whereas cuprous ions are assumed to contain a complete three-quantum group of 18 electrons, the same group in cupric ions would only contain 17 electrons, and the inner atomic structure would thus be asymmetric. Actually, cuprous ions are colourless and cupric ions coloured.

[5] Ladenburg had, in 1920, already endeavoured before Bohr to ascribe the chemical and magnetic properties of the elements enclosed in the rectangles in Fig. 47 to the formation of an "intermediate shell" of loosely bound electrons.

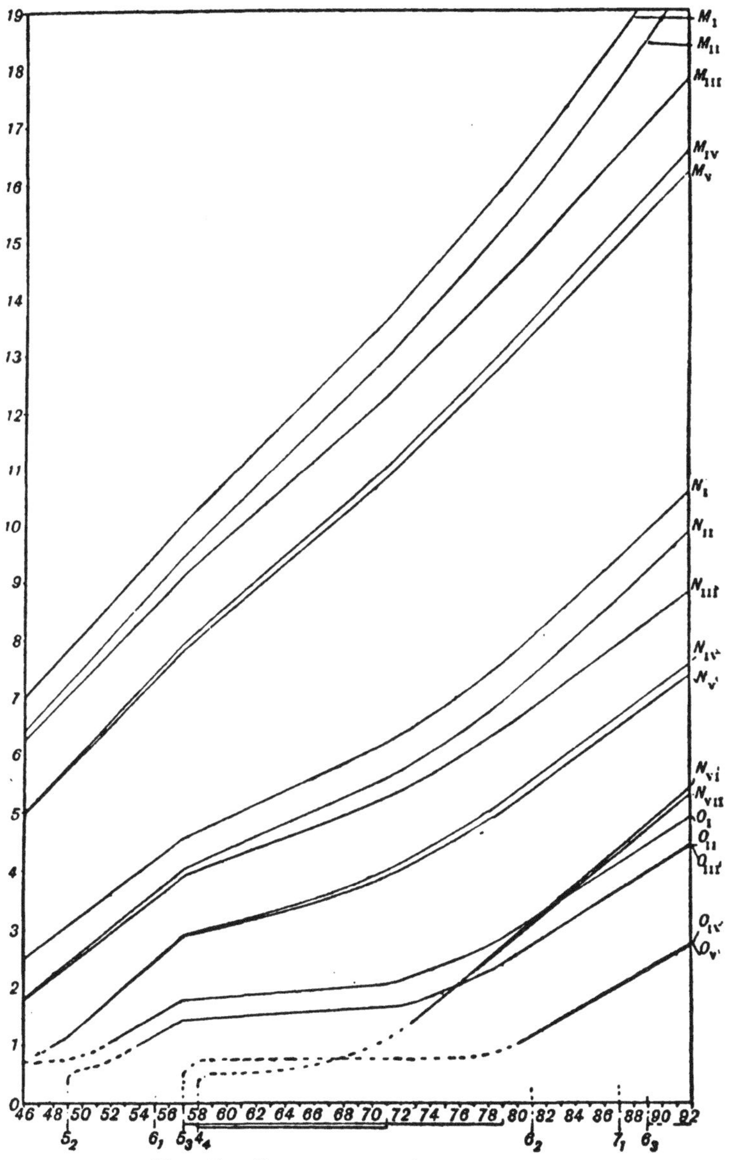

FIG. 48.—Röntgen terms of the elements.

in the same group has, at least, a partial screening effect. The completion of the three- and four-quantum groups from 8 to 18 electrons is also rendered evident by the smaller inclination of the curves.

In Fig. 48 a further remarkable fact is manifest. The 4_4-term is at first smaller than the 5_1- and 5_2-terms, and at the very beginning it is even smaller than the 5_3-term. As soon as the development of the four-quantum group of electrons is completed [at 71 (Lu)], however, the binding energy of the 4_4-term rapidly increases, so that, between the 75th and the 83rd element, the 4_4-curves *cross* the 5_1- and 5_2-curves.

§ 36. Fluorescence.

The radiation emitted by a body whose atoms are excited by primary rays is called *fluorescent radiation*. In the special case in which the wave-lengths of the secondary and the primary radiation are identical, we speak of a *resonance radiation*. According to the discussion in a previous section (§ 120), we may expect to meet with such radiation when light of the wave-length of a *resonance line* of an element passes through the vapour of the element at a temperature below that at which any of its atoms are excited.

The phenomenon of resonance radiation was discovered and investigated long before the atomic theory was established, first by *Wood* in the case of *mercury*, and subsequently by *Wiedemann*, *Dunoyer*, *Wood*, and *Strutt* in the case of alkali metal vapours, especially *sodium vapour*. At low pressures, the total energy of the re-emitted light does, in fact, agree with the energy absorbed from the primary radiation. It is only when the pressure is increased that a part of the absorbed primary radiation is changed into molecular energy of translation, with a resulting heating of the fluorescent vapour.

When a gas is irradiated with light of a frequency *greater than the resonance frequency* and corresponding to the difference in energy between the normal state and a *higher*, excited state, the *return* of the atom from this more highly excited state may take place *in steps*. In the fluorescent radiation, therefore, lines of greater wave-length than that of the exciting light may occur. Under no circumstances, however, can an atom which, before excitation, is in the normal state emit fluorescent radiation of shorter wave-length than the primary radiation, and in this lies the explanation of *Stokes's rule of fluorescence*, which was mentioned in an earlier section (§ 5). Only when the irradiated atom has been previously excited can it happen, by way of exception (*e.g.*, on a return to the normal state), that the wave-length of the fluorescent radiation may be shorter than that of the exciting radiation.

An especially clear example of the phenomena of fluorescence is afforded in the case of *mercury vapour*, as has been shown by *Füchtbauer*.[1] In his researches, a quartz mercury vapour lamp served as the source of the primary rays. Of the characteristic rays of mercury produced within the lamp, only those wave-lengths as long as, or longer than, the resonance wave-length of 2,537 Å.U. can pass through the quartz walls, the remainder being absorbed. In the path of the rays from the lamp is placed a tube containing cold mercury vapour. Since the atoms in the cold vapour may be assumed to be initially in the normal state, only rays of the resonance wave-length of 2,537 Å.U. will be absorbed, because any rays of greater wave-length present are incapable of raising the atom from its normal state to a higher one.[2]

The atoms which absorb the rays of 2,537 Å.U. wave-length are raised to the excited $2p_1$-state. The atoms thus excited are then able to absorb rays of 4,359 and 3,132 Å.U. wave-length as well from the primary radiation, whereby they are raised to the still higher states $1s$ and $3D$ (*cf.* Fig. 45 in § 33). Now, in accordance with the selection principle (§ 33), the mercury atom can undergo transitions from the $1s$-state to the $2p_2$- or $2p_0$-states, with the emission of the lines 5,461 or 4,047. Again, the selection principle permits of transitions from the state $3D$ to $2P$ and $2p_2$, with the emission of the lines 5,791 and 3,663. All these lines could actually be observed by Füchtbauer in the fluorescence spectrum; and it was evident that they were all produced by rays of 2,537 Å.U., because they all vanished when the latter were absorbed by a thin glass plate. At the same time, we see from Fig. 36 how the transition from the $2P$- to the normal $1S$-state (indicated by the dotted line) gives rise to an indirect fluorescent radiation of *shorter wave-length* than the primary radiation. Furthermore, Fig. 36 clearly shows how the primary light, which, being ultra-violet, is not visible

[1] *Physikal. Zeitschr.*, 21 (1920), p. 635.

[2] The other resonance line of mercury, of 1,849 Å.U wave-length, which is due to a transition on absorption from the normal to the $2P$-state, is not found in the primary radiation because rays of this wave-length are absorbed by the quartz.

to the human eye, gives rise as a result of fluorescence to a succession of rays whose wave-lengths lie between 4,000 and 8,000 Å.U., and are thus visible.

Just as atoms can be excited otherwise than by absorption of radiation, so also we must assume that an atom excited in any way can *return* to its normal state *without radiating.* In particular, we may expect the excess energy to be given off without radiation when the excited atom *collides* with another atom ; in that case, the excitation energy will be *transformed* into *translational energy*, with a consequent increase in the heat energy present.

Thus, of the excited atoms in a fluorescent gas, only a certain number would, in general, return to the normal state with the emission of radiation, the others returning without radiating. The ratio between the two numbers, however, evidently depends on the ratio between the *mean time between two collisions* (to be calculated from the theory of gases) and the *average duration of the excited state.* As long as the former interval is considerably greater than the latter, the total energy of the fluorescent radiation must be approximately as large as the total energy absorbed from the primary radiation by the fluorescing gas. If, however, the mean interval between two collisions be *shortened*, as by *increasing the pressure* by the *admixture of another gas*, the *fluorescence* should be gradually *extinguished* as the mean interval between two collisions becomes of the same order of magnitude as the average duration of the excited state.

The extinction of fluorescence was actually observed by *Wood*, when air was mixed with the fluorescent mercury vapour.[3] Both his observations and those of later investigators [4] lead to an approximate value of 10^{-8} sec. for the *mean duration of the excited state.* One-tenth as large (as we may remark in passing) is the time found by *Wien* for the *decay of the luminescence* of excited canal ray particles.[5]

Moreover, it has been found possible to demonstrate experimentally a *transference of excitation energy* from the atoms of one substance to those of *another*, the transference being obviously effected by collisions. This phenomenon, in which light is absorbed by atoms of one kind and re-emitted by atoms of another kind, is known as *sensitized fluorescence.*

[3] *Phil. Mag.*, **21** (1911), p. 309.
[4] *Cf.* G. Cario, *Zeitschr. f. Phys.*, **10** (1922), p. 185.
[5] W. Wien, *Ann. der Phys.*, **60** (1919), p. 597, and **66** (1921), p. 229.—The previously discussed experiments of Füchtbauer also led to a probable value of 10^{-8} sec. for the average duration of excitation.

It was first observed, in 1922, by *Franck* and *Cario*,[6] in a *mixture of mercury vapour and thallium vapour.* This mixture was irradiated at suitable pressure and low temperature with ultra-violet mercury light of the resonance wave-length of 2,537 Å.U. These rays are naturally absorbed by the mercury atoms, but not by the atoms of thallium, whose spectrum exhibits in the neighbourhood of 2,500 Å.U. no lines having the normal state as initial state for absorption. None the less, the spectrum of the fluorescent radiation contained all the lines of thallium whose excitation energy is smaller than a light-quantum of 2,537 Å.U. wave-length. Similar experiments were performed with mixtures of mercury and lead, mercury and bismuth, mercury and silver, and mercury and cadmium. At higher temperatures, indeed, sensitized fluorescence makes its appearance also with lines whose excitation energy is larger than a light-quantum of the primary radiation. In such cases the deficit in energy is evidently supplied during collisions by the translational energy of the mixture.

§ 37. Electron Collisions.

As we have seen in the previous considerations, Bohr's theory of spectra and atomic structure is based on the blending of two fundamental hypotheses. One of these is that the internal energy of atoms can assume only quite definite and discrete values; the other finds expression in Bohr's frequency condition. The theoretical explanation of *spectral* phenomena is to be found in a *combination* of these two fundamental assumptions.

It was, therefore, a great advance for atomic physics when, in 1913, *Franck* and *Hertz* succeeded in effecting a *direct experimental confirmation* of the first of the two fundamental hypotheses *quite independently* of the second. They were able to do this by investigating *collisions between atoms and free electrons* which, in the form of slow negative corpuscular rays, were made to pass through a gas or a vapour at a sufficiently low temperature. The phenomena of so-called *electron collisions*, as they are briefly termed, must for two reasons obviously afford much simpler and clearer

[6] *Cf.* the summary by J. Franck, "*Neuere Erfahrungen über quantenhaften Energieaustausch bei Zusammenstössen von Atomen und Molekülen,*" in the second volume of "*Ergebnisse der exakten Naturwissenschaften*" (Berlin: Springer, 1923).

experimental results than collisions between atoms. In the first place, the kinetic energy of a single electron can be exactly determined from a measurement of the electrostatic potential. On the other hand, the mass of an electron is negligible compared with that of an atom, and hence the exactly measurable kinetic energy lost by an electron can at once be equated to the increase in internal energy experienced by the atom involved in the collision.

Now, as long as the kinetic energy of an electron is smaller than the energy required to excite an atom, the kinetic energies of the electrons will not be diminished by the collisions. In this case, *only "elastic" impacts* are possible. *Inelastic impacts* (in which the colliding electron loses energy) will first be possible when a critical value of the potential is reached, such that, when multiplied by the elementary quantum of electricity, the product is equal to the excitation energy; this critical value is known as the *excitation potential.*[1] If we denote by λ^* the wave-length of the line emitted by the atom in its return from the excited state (in the narrower sense of the word) to the fundamental state, and by V^* the excitation potential, we have, by § 35 eqn. (1), the relation

$$(1) \qquad \lambda^* V^* = 1{\cdot}234 \times 10^4 \ (\lambda^* \text{ in Å.U., } V^* \text{ in volts}).$$

For the elements in the first group of the periodic system, λ^* is to be given the value of the wave-length of the *resonance line*. With the *elements in the second group*, for which there are two resonance lines differing considerably in wave-length, a corresponding distinction is made between *two different excitation potentials*.

For many elements, it has been possible to make direct *experimental determinations* of the excitation potential by purely electrical methods. Using suitable apparatus, the electrons are made to pass through the gas or vapour, and the resulting current is measured with a galvanometer, while the applied potential difference is also observed. If the current be plotted on a graph against the potential, the curve exhibits a strongly marked discontinuity for the potential at which inelastic impacts first occur. With many elements, the *first flashing out of the resonance line* caused by the electronic collisions could be observed at this potential,

[1] Obviously, however, the kinetic energy of electrons can be increased before the excitation potential is reached, if they collide with previously excited atoms, which then give up their excess energy; in that case we speak of "electron collisions of the second kind."

and this in itself must be regarded as a brilliant experimental confirmation of Bohr's frequency condition.[2]

From the point of view of Bohr's theory, further discontinuities in the current are to be expected for values of the potential which, when multiplied by the elementary quantum of electricity, give the energy difference between the fundamental state and a state excited to a higher degree. The

TABLE XVI

Excitation and Ionization Potentials (in Volts)

A. ALKALI METALS.

		Li	Na	K	Rb	Cs
Excitation potential	calc.	1·84	2·09	1·61	1·6	1·4
	obs.	—	2·12	1·55	1·6	1·48
Ionization potential	calc.	5·37	5·12	4·32	4·16	3·88
	obs.	—	5·13	4·1	4·1	3·9

B. METALS OF THE SECOND GROUP.

		Mg	Ca	Zn	Cd	Hg
Excitation potential (I)	calc.	2·7	1·88	4·01	3·78	4·87
	obs.	2·65	1·90	4·18	3·95	4·76 (4·9)
Excitation potential (II)	calc.	4·33	2·92	5·77	5·39	6·67
	obs.	4·42	2·85	5·65	5·35	6·45 (6·7)
Ionization potential	calc.	7·61	6·09	9·35	8·95	10·39
	obs.	7·75	6·01	9·3	8·92	10·2

determination of these closely adjacent potential values, however, generally meets with experimental difficulties, and has hitherto only been successful in the case of mercury.[3] On the other hand, the potential at which the electron impacts first produce ionization of the atoms of the gas can be fairly sharply determined for many elements. This

[2] Thus, in mercury, Franck and Hertz observed the first inelastic impact at 4·9 volts, and at this same potential the first occurrence of the resonance line 2,537 Å.U.

[3] Franck and Einsporn, *Zeitschr. f. Phys.*, 2 (1920), p. 18.

potential is called the *ionization potential.* It must be equivalent to the *fundamental term* of the atom, inasmuch as this corresponds to the work necessary for the removal of the electron concerned from the atom. The ionization potential must therefore satisfy eqn. (1) for a wave-length corresponding to the fundamental term.

It has been possible to *test* eqn. (1) *experimentally* for the *alkali metals* and the *metals of the second group* of the periodic system, both as regards the excitation potential and the ionization potential. Table XVI gives a summary of the calculated and the observed potential values. The calculations were carried out with the help of eqn. (1), taking, in the case of the alkali metals, the wave-length of the first line of the principal series ($2p \rightarrow 1s$) and the $1s$-fundamental term, and, for the metals of the second group, the wave-lengths of the two resonance lines and the $1S$-fundamental term. The observed values of the potential are those at which the first inelastic electron impacts and the first occurrence of ionization were actually observed. As Table XVI shows, a very good agreement exists between the calculated and the observed values.[4]

The excitation and ionization potentials have been measured for numerous elements in the other groups, apart from the elements in the first two groups of the periodic system. Thus, for example, the excitation potentials for thallium and lead have been found to be 1·07 and 1·26 volts respectively, and the corresponding ionization potentials 7·3 and 7·93 volts. By means of the method of electron collisions, it was in this way possible to derive two optical terms for lead, although a resolution of its spectrum into series has not yet been effected.

For the elements of the second group of the periodic system, and for some of the rare gases, the *ionization potentials of the second order* have also been observed, *i.e.*, the potentials at which a second electron is removed from the atom. Just as the ionization potential of the first order is equivalent to the fundamental term of the arc spectrum, so the ionization potential of the second order corresponds to the *fundamental term of the spark spectrum.* The ionization potential of the second order is naturally considerably higher than that of the first order; *e.g.*, for magnesium, the two potentials

[4] The observed values given are those of *Foote* and *Mohler*, who have developed Franck and Hertz's method. *Cf.* Foote and Mohler's "*The Origin of Spectra*" (New York, 1922). The bracketed values for Hg in Table XVI are based on measurements by Franck and Einsporn (*cf.* note 3).

amount to 7·6 and 15 volts, and for helium to 24·5 and 78·6 volts.[5]

§ 38. Metastable Atomic States.

As we have seen in an earlier section, an atom generally persists in an excited state for an interval which is found to be only 10^{-8} sec. in order of magnitude; provided that the excited state has not already been destroyed, the atom then returns *spontaneously* to its normal state, with the *emission of monochromatic radiation.* It is, however, conceivable that an atom might be raised to a state of higher quantum number than the normal by some *other means* than the absorption of radiation, from which state a spontaneous return, with the emission of radiation, to the normal state would be *precluded by the spectroscopic selection principle.* In that case, the atom would have to persist in the excited state until this was *destroyed* by some *external influence*, such as collisions with other atoms. Thus, it appears theoretically possible for an abnormal state of this kind to continue for an arbitrary length of time.[1] Such an excited state, from which there is *no spontaneous return to the normal state*, is termed *metastable.*

Metastability has actually been detected in *atoms of helium.* In the arc spectrum of *neutral* helium [2] there are *two different systems of terms*, between which *no spectroscopic transitions* occur. The two systems of terms and series are distinguished by two names which are now only of historical significance ; *viz.*, *parhelium* and *orthohelium.*[3] The normal orbit of parhelium, which corresponds to the normal state of the helium atom, must be assumed to be a one-quantum orbit, whereas the lowest orbit of orthohelium must be a

[5] The frequency which, according to eqn. (1), corresponds to the ionization potential of the first order (24·5 volts) for helium must evidently represent the K-term of helium, since, quite generally, the K-term is a measure of the work required to liberate an electron from a normal one-quantum orbit. A comparison of the K-term of helium thus derived with the known K-terms of Be, B, C, N, and O (*cf.* § 30), and of the elements from Na onwards, clearly shows that from helium to sodium the natural sequence of the elements completely agrees with the series formed in order of increasing atomic weights. The nuclear charge numbers of all the elements are thus unequivocally determined, for a comparison of the spectra of ionized helium and hydrogen immediately shows that the nuclear charge number of helium is two.

[1] Provided only that pressure and temperature be kept sufficiently low.

[2] We must, of course, distinguish between neutral helium and the hydrogen-analogue, ionized helium.

[3] The series of orthohelium consist of very narrow doublets, whereas the lines of the parhelium spectrum are singlets. The well-known strong yellow helium line (the Fraunhofer line D_3) belongs to orthohelium ; its term representation is $2p - 3d$.

two-quantum one. As we have just said, no spectroscopic transition between the normal para-term and the lowest (two-quantum) ortho-term is possible. By means of *electron collisions*, however, *Franck* has succeeded in transforming helium atoms from the normal para-state into the lowest ortho-state, for which process a so-called *transformation potential* of 19·8 volts is necessary. The actual metastability of the two-quantum ortho-state thus produced follows from the fact that this state is associated with a characteristic *resonance line*. Its term representation is $2_2 \rightarrow 2_1$ (both terms, of course, being ortho-terms), and its wave-length is 10,830 Å.U., or approximately 1μ[4]. For, if helium in which a sufficient number of metastable atoms has been produced by electron collisions be irradiated with infra-red light of about 1μ wave-length, perfect resonance ensues. This clearly shows that, in spectroscopic respects, the 2_1-ortho-state represents a normal state from which there is no spontaneous return, with the emission of radiation, to the true normal state (*viz.*, the normal para-state).[5]

The metastable state is, of course, associated with a *smaller ionization energy* than the normal state. The difference between the ionization potentials of the normal and the metastable atom must, indeed, be equal to the transformation potential. Since the ionization potential of normal helium is 24·5 volts, and the transformation potential 19·8 volts, the ionization potential of metastable helium must amount to 4·7 volts. The metastable helium atom must therefore lose an electron more easily than lithium and sodium,[6] and hence must be still *more electropositive* than these alkali metals, and, in contrast to the rare gases, it should exhibit considerable *chemical activity*. This also follows from the fact that, at sufficiently high densities, a band spectrum of helium appears, and this is obviously due to the *formation of molecules* by metastable atoms.[7]

[4] As was mentioned in the prefatory note, a micron (μ) is one-thousandth of a millimetre.

[5] If, on the other hand, helium be irradiated with infra-red light of about 2μ wave-length, only incomplete resonance results, for in the spectrum of parhelium there is a line which has approximately this wave-length (to be more exact, 20,582 Å.U.); its term representation is $2S - 2P$ (the para-terms being distinguished by capital letters). This line is absorbed by such atoms as have been *previously* excited from the normal state $1S$ into the para-state $2S$. After absorption of the 2μ line, the atoms thus still further excited can now return to the normal state with the emission of radiation, so that only some of these atoms emit a pure resonance radiation.

[6] *Cf.* Table XVI in § 37.

[7] According to two notes in "*Nature*" [114 (1924), p. 861, and 115 (1925),

§ 39. Atomic Magnetism.

Since every revolving electron in an atom represents a convection current whose strength is given by the product of the elementary quantum of electricity and the revolution frequency (ω), every *electron orbit* in an atom is associated [according to a well-known formula of electrical theory] with a *magnetic moment* of magnitude

$$(1) \qquad M = \frac{e\,\omega\,F}{c}$$

where F denotes the area of the orbit.

Let us consider, in particular, the *normal hydrogen atom*, for whose circular orbit (of radius a) the angular momentum (U) is equal to $h/2\pi$; then

$$(2) \qquad U = \frac{h}{2\pi} = m\,a\,v = m\,a\,.\,2\,\pi\,a\,\omega = 2\,m\,\omega\,F\,.$$

Hence, denoting the specific charge of an electron by γ, we have

$$(3) \qquad M^* = \frac{U\gamma}{2c} = \frac{h\gamma}{4\pi c}\,.$$

Introducing the known values of h and γ, we find

$$(4) \qquad M^* = 9{\cdot}21 \times 10^{-21} \text{ abs. units.}$$

The quantity M^* is known as *Bohr's magneton*.

Although every electron orbit is associated with a magnetic moment, yet an atom as a whole may be non-magnetic, owing to a mutual *compensation* of the magnetic moments of the electronic orbits. This will be the case when two parallel orbits, of equal area and with the same revolution frequency, are traversed in opposite senses. A compensation of this kind must, however, become imperfect when the previously non-magnetic atom is exposed to the action of an external *magnetic field*. For, in every circuit, an *induction current* will then be produced, and the sense of this current will depend only on whether the magnetic flux of force traversing the circuit area increases or decreases. Thus, although the original, mutually compensating, currents may be opposed in sense, yet the superimposed induced currents will be *similar in sense*, so that the original compensation will be partially destroyed. Since no resistance is offered to the induced currents, they must persist until, on removal of the magnetic field, they disappear as a result of a second induc-

p. 337] J. J. Manley has succeeded in combining mercury and helium under the influence of a glow discharge to form a compound HgHe.

tion effect opposed to the first one. In a magnetic field, a non-magnetic atom therefore exhibits *induced magnetism.* According to Lenz's rule, however, a *force of repulsion* must exist between the induced currents, when they are generated by the approach of a magnet, and the inducing magnet. Consequently, the substance in which the induced currents are flowing must appear *diamagnetic*.

Atomic theory regards as *paramagnetic* those substances in which, because of an imperfect compensation (*cf.* § 35) in the atoms (or molecules), the vectorial sum of the magnetic moments differs from zero from the very beginning, whilst an external magnetic field exerts *directing forces* on the atoms. In *ferromagnetic* bodies, which exhibit a permanent magnetism, there appear to be, in addition, *molecular forces* acting between the molecules, these forces ceasing to act, however, above a certain temperature, known as the *Curie point*. Above this temperature the body in question exhibits only the properties of paramagnetism (in the narrower sense); *e.g.*, for iron, the Curie point lies at 750°, and for nickel at 375° C.[1]

In 1911, *Pierre Weiss* came to the conclusion that, in a state of magnetic *saturation* (only attainable at low temperatures, because at higher temperatures the continuous molecular collisions do not admit of it),[2] the *magnetic moment* referred to a gram-molecule is, in ferromagnetic substances, an *integral multiple* of 1,123 abs. units. For example, below the Curie point it is three times that amount in nickel, nine times in cobalt, and ten or eleven times in iron. The magnetic moment of a single molecule would thus be an integral multiple of the amount obtained on dividing 1,123 by the Loschmidt number, *i.e.*, of

$$M' = 1{\cdot}85 \times 10^{-21} \text{ abs. units.} \tag{5}$$

This quantity, which is known as *Weiss's magneton*, is one-fifth as large as Bohr's magneton. More recent investigations (which we cannot discuss here), however, have explained the apparent discrepancy in favour of the quantum theory.

The concept of atomic magnetism dates back for more than a century. It is based principally on the well-known experimental fact that, even in the smallest fragments of a magnet, the algebraical sum of the quantities of magnetism contained therein vanishes. On the analogy of the equiva-

[1] At the Curie point a sudden change in the specific heat also occurs.

[2] In the state of saturation the vectors of the magnetic moment are presumably similarly directed in all molecules.

lence between closed currents and magnetic shells which he had discovered, *Ampère* (1822) referred atomic magnetism to electric currents circulating in the individual molecules. A further important support for this concept was provided by *Faraday's* discovery (1845) that magnetism is a universal property of all matter.[3]

The *reality* of the so-called *molecular currents* conjectured in theory was first demonstrated in 1915 by *direct experiment* by *Einstein* and *de Haas*, who succeeded in causing a soft iron rod to *rotate* by *rapid reversals of magnetism*.[4] An effect of this kind could, indeed, be predicted on the basis of eqn. (3), according to which every change in the magnetic moment of an atom connotes a change in its angular momentum, so that, in virtue of the Principle of the *Conservation* of the Total Angular Momentum, a jerky rotation of the whole body must result as a *reaction effect*. In 1917, *Barnett* succeeded in producing the converse of the effect discovered by Einstein and de Haas, *viz.*, in *magnetizing a rod of soft iron by rapid rotation*.[5] It may be remarked incidentally that the results of observations on both effects give a value for the specific charge of an electron which is exactly half that obtained from deflection experiments on cathode rays.[6]

An important supplement to Einstein's and Barnett's experiments was furnished by *Gerlach* and *Stern* in 1922, when they passed a beam of rapidly moving *atoms of silver* through a *non-homogeneous magnetic field*. (A non-homogeneous magnetic field exerts upon a magnet placed in it not only directing, but also deflecting, forces.) This experiment enabled Gerlach and Stern to show that, just as was demanded by the *quantum theory* in opposition to the classical theory, only *quite definite orientations* of the atoms relative to the magnetic field are possible.[7] The value of the magnetic

[3] The above-given explanation of diamagnetism from the point of view of the molecular current hypothesis is essentially that of *Wilhelm Weber* (1852). The development of Weber's ideas on an electron-theoretical basis is principally due to *Gans* and *Langevin*.

[4] *Verhandl. der Deutsch. Physikal. Gesellsch.*, **17** (1915), p. 152.

[5] *Phys. Rev.*, **10** (1917), p. 7.

[6] The theoretical explanation of the factor 1/2 is connected with similar difficulties in the interpretation of the so-called anomalous Zeeman effect, which are not yet solved.

[7] As Debye and Sommerfeld have shown, it follows from the quantum theory that an n-quantum atom can only assume $2n$ positions in a magnetic field. The angle (ϕ) included between the normal to the orbital plane and the magnetic lines of force is given by the relation

$$\cos\phi = \pm\frac{n'}{n}\,(n' = 1\ 2\ \ldots\ n).$$

moment of a single silver atom was found from the experiment to be actually that of Bohr's magneton.[8]

The greatest difficulty with which the older theory of atomic magnetism had to contend was the question as to how *resistance-free circuits* could possibly exist.[9] The modern theory, it is true, is also not in a position to solve this difficulty, but it recognizes in it only a special case of a much more general difficulty, which arises from a *contradiction* between *Bohr's atomic theory* and *Maxwell's theory* of the electromagnetic field. It follows from *Poynting's theorem*, which is a necessary consequence of Maxwell's equations, and from the concept of convection currents, that every *electric charge* when *accelerated* in any way must *radiate electromagnetic energy*. According to Poynting's theorem, the same must obviously also hold for every electron revolving within an atom, in which case it would no longer be possible for the electrons to describe stationary orbits as postulated by Bohr's atomic theory. A key to the removal of this contradiction, however, already appears to be provided in the so-called correspondence principle of Bohr.[10]

§ 40. The Formation of Molecules.

Our previous considerations have shown that an atom, on *losing* any of its loosely bound electrons, becomes a *positive ion ;* on the other hand, it is transformed into a *negative ion* when the number of electrons surrounding the nucleus is *raised* above the nuclear charge number. The actual existence of such negative ions is rendered evident by the phenomena of *electrolysis ;* halogens are electrolytically separated as univalent negative ions, elements of the sixth group of the periodic system as divalent negative ions, and so on.

The formation of negative ions is presumably to be explained by ascribing to *neutral atoms* of certain elements a definite *affinity for electrons*, *i.e.*, an ability to *combine* with free electrons, with the *liberation of energy*. Just as energy is liberated when a positive ion recombines with an electron

[8] For further details of Gerlach and Stern's researches, as well as of the problems discussed in this section, *cf.* W. Gerlach's report, " *Magnetismus und Atombau* " in the second volume of " *Ergebnisse der exakten Naturwissenschaften.*"

[9] At extremely low temperatures, moreover, a number of metals become what are termed *supraconductors*, and in them *Kamerlingh Onnes* and his collaborators have been able to produce currents practically without resistance. *Cf.* the report by Crommelin in the *Jahrb. f. Radioakt. u. Elektronik*, **19** (1922), p. 38.

[10] *Cf.* § 12.

previously removed from the atom,[1] so also must energy be liberated when an atom which is neutral, but has an affinity for electrons, takes up an electron.

An empirical confirmation of this conception is apparently afforded, as *Franck* has shown,[2] by *spectroscopic* experiments which permit of a quantitative determination of the electron affinity by optical means, *i.e.*, a determination of the energy liberated in the formation of negative ions. For, under certain circumstances, this energy can be transformed into electromagnetic radiation energy, and a light-quantum thereby generated must obviously be equal to the sum of the energy liberated on the formation of an ion and the kinetic energy possessed by the electron before its binding. Since the kinetic energy may have any arbitrary value, which, however, must always be positive, we may expect the appearance of a *continuous spectral band with a limit on the side of longer wave-lengths*, and the frequency of this limit multiplied by h must be a measure of the electron affinity.

Since *oppositely charged* ions exert an *electrostatic force of attraction* upon each other, the combination of ions may lead to the formation of a *molecule of a chemical compound*. Observations of the *evolution of heat* accompanying chemical processes give us an idea, correct at least in order of magnitude, of the amount of *energy liberated* in the formation of such a compound. It must not be forgotten, however, that alterations in the state of aggregation frequently find expression in the observed evolutions of heat.[3] In the most usual chemical reactions, the evolution of heat referred to the gram-molecule amounts to between 10,000 and 100,000 calories,[4] *i.e.*, to between 4×10^{11} and 4×10^{12} ergs. On dividing by the Loschmidt number, we obtain the amount associated with a single molecule—approximately 7×10^{-13} to 7×10^{-12} ergs. By way of comparison, it may be remembered that the excitation energies of the alkali metals lie between 2×10^{-12} ergs and one and a half times that amount.

Up to now we have limited our considerations to molecules formed by the union of oppositely charged ions; such compounds are termed *heteropolar*. It is, however, possible for two neutral atoms to exert a force of attraction on each

[1] This energy is, of course, identical with the ionization energy.

[2] *Ann. der. Phys.*, 64 (1921), p. 673.

[3] Therein also lies the explanation of the observation of apparently negative evolutions of heat.

[4] The heat evolved, for example, in the formation of one gram-molecule of water from H_2 and O amounts to 69,000 cal., and in the formation of CO_2 from C and O_2 to 96,000 cal.

other, provided that the electrons in the one atom come sufficiently near the protons in the other; in that case, *homopolar* compounds are formed.

The most important examples of homopolar compounds are represented by those formed *from atoms of the same element.* It follows from measurements of the gas density (and from Avogadro's Law [5]) that, in the gaseous state, the molecules of *hydrogen, nitrogen, oxygen,* and the *halogens* are all *diatomic,* as are those of selenium and tellurium. Besides diatomic molecules, molecules of *eight atoms* have been found in *sulphur,* and of *four atoms* in *phosphorus* and *arsenic,* below certain limits of temperature.[6] The modification of oxygen known as *ozone* has been proved to be *triatomic.* Determinations of the vapour densities of the electropositive metals have invariably shown that these do not form molecules, but are, chemically speaking, *monatomic.* That such is also the case with the rare gases in general is evident from their chemical inertia.[7] Homopolar combinations between whole molecules are also obviously possible; and in this way arise the so-called *complex compounds* in which whole molecules appear to group together.[8]

The concept of the *electrostatic nature* of the *chemical forces* acting between atoms was suggested by *Berzelius* in 1812. Since, however, both homopolar compounds and complex compounds, as they gradually became known, appeared to contradict Berzelius' theory, it dropped into the background after ruling chemistry for several decades. It was not until the existence of both positive and negative

[5] According to Avogadro's Law, equal volumes of all gases contain, at the same pressure and the same temperature, equal numbers of molecules.

[6] At higher temperatures, both phosphorus and arsenic vapours are apparently diatomic.

[7] The assumption of a formation of molecules in helium under certain circumstances has already been discussed in § 38.

[8] For example, both the ammonia molecule (NH_3) and the hydrochloric acid molecule (HCl) are, as a whole, neutral. Nevertheless, in consequence of the attraction exerted by the nitrogen ion, the NH_3 molecule is able to combine with the hydrogen ion in the HCl molecule, thereby forming the complex compound ammonium chloride (NH_4Cl). Indeed, the binding between the two molecules is so strong that, even in electrolysis, ammonium chloride behaves like a compound of a negative chlorine ion and a univalent cation NH_4, known as "ammonium," which in its combinations plays a precisely similar part to the likewise univalent ions of the alkali metals. The chlorine atom in the HCl molecule, like the hydrogen atom, can also join on to other molecules. The number of atoms contained in the molecules of organic compounds can be very large, *e.g.*, a molecule of cane-sugar consists of 45 atoms (corresponding to the formula $C_{12}H_{22}O_{11}$). A molecule of hæmoglobin, the red colouring matter in blood, is probably composed of 2,378 atoms, corresponding to the hypothetical formula $C_{758}H_{1203}N_{195}O_{218}FeS_3$.

charges in every atom was recently established that *Kossel*, in 1916, was able to revive Berzelius' theory by explaining the contradiction.

By the *absorption of light-quanta*, molecules can be brought into a state in which, because of their *increased energy*, they are enabled to *take part in chemical reactions* not possible for them in their normal state. The phenomena based on this absorption are termed *photochemical*. In 1912, *Einstein* formulated his *law of photochemical equivalence*, which states that the number of chemically reacting molecules must be equal to the number of absorbed light-quanta. Experimental measurements seem to confirm this principle.[9] Among the most important and best-known photochemical reactions are the formation of hydrochloric acid (HCl) from H_2 and Cl_2, the assimilation of carbon dioxide by plants, and the decomposition of the silver halides upon which photography is based.

Like the phenomena of fluorescence, photochemical reactions can also be "*sensitized*." That is to say, under certain circumstances the reactions can also be produced by light of wave-lengths which are not absorbed by the reacting molecules, provided only that a sensitizer which can absorb the light be added. Owing, presumably, to molecular collisions, the surplus energy absorbed is transferred from this sensitizer to the molecules of the substances concerned, which can then react. This appears especially clearly from an experiment of *Franck* and *Cario*, in which *mercury vapour* served as sensitizer for the *dissociation of hydrogen*.[10] It is known from thermochemical measurements that the energy required for the resolution of a molecule of hydrogen into its two atoms amounts to about 6×10^{-12} ergs. Dividing this by the elementary quantum of action, we obtain a frequency of about 9×10^{14} sec.$^{-1}$, or a wave-length of approximately 3,000 Å.U.[11] A non-excited hydrogen atom, however, cannot absorb any radiation of greater wave-length than its resonance line, which (as the first line of the Lyman series) lies at 1,216 Å.U., far in the ultra-violet. Franck and Cario introduced mercury vapour into the hydrogen, and

[9] As a result of the frequent occurrences of intermediate reactions, most photochemical reactions appear very complicated.

[10] Cario and Franck, *Zeitschr. f. Phys.*, 11 (1922), p. 161.

[11] The corresponding potential is 3·5 volts. This "dissociation potential" actually represents the excess potential necessary for the generation of definite hydrogen lines over and above the values equivalent to the initial levels on emission. From this we must conclude that a hydrogen molecule must be split up into atoms before excitation can take place.

submitted the mixture at sufficiently low temperature to the action of the rays of 2,537 Å.U. wave-length emitted by a quartz mercury lamp. Although, in accordance with what we have just said, these rays could not be absorbed by the hydrogen, yet a dissociation of the hydrogen took place. This was clearly proved by a reduction of metallic oxides, which, as is known from experience, can only be effected at the temperature in question by monatomic, but not by diatomic, hydrogen. For the rest, the best-known instance of a photochemical sensitization is afforded by the *sensitization of a photographic plate* with suitable dyes which possess in the red the absorption lines lacking, in that region, in the silver halides.[12]

The converse of photochemical processes is represented by *chemico-luminescence*, due to a transformation of chemical energy into light-quanta. A well-known example of this is the oxidation of aldehydes at low temperatures, which is accompanied by an intensive luminescence. *Haber* and *Zisch* have shown that, on leading a stream of chlorine through sodium vapour, chemico-luminescence ensues; it can be recognized by the appearance of the characteristic *D*-lines of sodium at a temperature at which a luminescence due to heat is obviously excluded.[13]

Chemical reactions may be produced not only by the absorption of light-quanta, but also by *electron collisions*. The experimental investigation of these processes promises to be of great importance, because it permits of an exact determination of the *minimum energy* requisite for the initiation of a chemical reaction.[14]

§ 41. The Structure of Crystals.

A special case of the formation of molecules is represented by the *formation of crystals*. The example of the rock-salt crystal, discussed in an earlier section, showed that crystals represent single *giant molecules*. On account of the absolutely regular arrangement in space-lattices, however, the

[12] Radiation absorbed by molecules can, however, be re-emitted by them in the form of fluorescent radiation. Hence it in no way follows that a molecule which absorbs an amount of energy sufficient for dissociation should necessarily be dissociated. At low pressures this hardly ever occurs. For example, diatomic iodine vapour, when excited by the green mercury line, radiates fluorescent light, although the energy of a green light-quantum is half as large again as the dissociation energy of the iodine molecule.

[13] Haber and Zisch, *Zeitschr. f. Phys.*, **9** (1922), p. 267.

[14] Buch-Andersen has investigated the formation of ammonia (NH_3) from nitrogen and hydrogen on these lines: *Zeitschr. f. Phys.*, **10** (1922), p. 54.

different species of atoms are intercombined in very simple numerical proportions, so that the formation of crystals may be associated with a simple chemical formula and, correspondingly, a definite "molecular weight."

As was explained in an earlier section (§ 19), the possibility of crystal analysis is based on the fact that atoms of different mass diffract Röntgen rays differently. Diffracting power depends, in fact, on the number of electrons surrounding the atomic nucleus, and this number increases in general with the atomic weight. Instead of the original Laue photographs, line spectra of known wave-lengths are now used for the investigation of crystal structure, whence, by means of Bragg's equation, the constants characteristic of the structure can be derived. Moreover, the ground

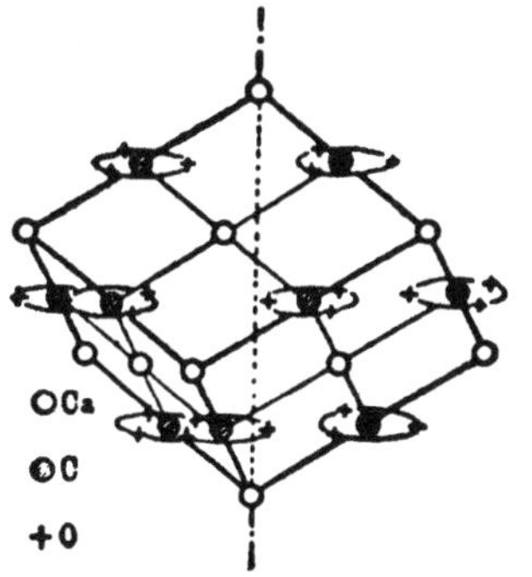

FIG. 49.—Structure of calcite.

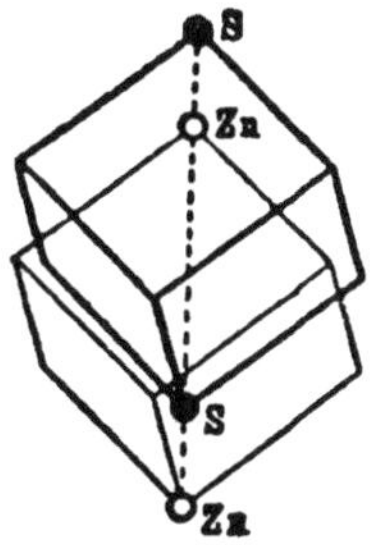

FIG. 50.—Structure of zinc blende.

available for research in crystal analysis has been greatly extended through the application of the crystal powder method (already mentioned in § 19).

Most *elements* crystallize in *face-centred cubic lattices*, such as is obtained by replacing the Na and Cl atoms in the rock-salt crystal (*cf.* the earlier Fig. 28) by atoms of the same kind. Crystal structures of this nature have, for example, been found in copper, silver, gold, calcium, aluminium, lead, cobalt, nickel, rhodium, palladium, iridium, platinum, and also in nickel-steel and manganese-steel. The lattice constants of these crystals lie between 3·5 and 5·5 $\times 10^{-8}$ cm. The very great *stability* of the face-centred cubic arrangement is proved by its persistence, with unchanged lattice constant, even when the metals are sprayed ; thus, particles consisting of only a few hundred atoms probably still have the same crystal structure. *Space-centred* cubic crystal lattices have been discovered in some metals, such as lithium, sodium, tantalum, chromium, molybdenum, tungsten, and iron. Lattices of the rock-salt type are also found in numerous compounds, *e.g.*, in KCl, KBr, KI, LiF, and in others. Crystals of cæsium chloride represent a related type to that of rock-salt; the Cs atoms by themselves form a simple cubic lattice, as do the Cl atoms, and the points in one lattice coincide with the centres of the cubes in the other.

Besides the rock-salt type previously discussed, the calcite ($CaCO_3$) type is also of great importance (Fig. 49). We can best picture to ourselves the transition from one type to the other by imagining the rock-salt lattice to be placed with a diagonal of the cube vertical, and the Na and Cl atoms to be

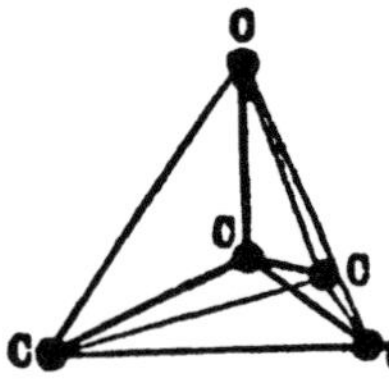

FIG. 51.—Structure of diamond.

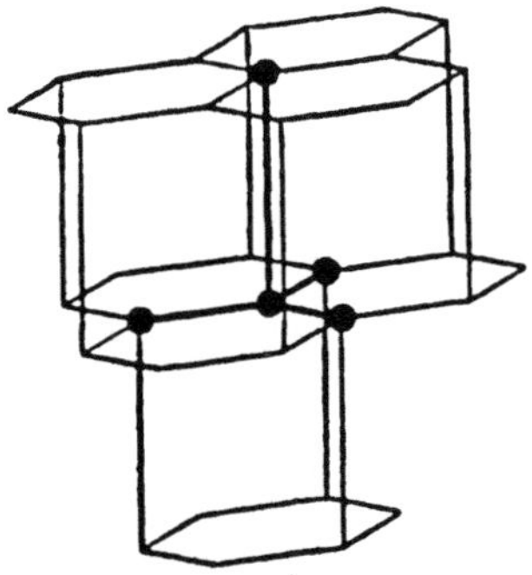

FIG. 52.—Structure of graphite.

replaced by Ca and C atoms. Then we suppose three O atoms to be added round each of the C atoms in a horizontal plane, *i.e.*, at right angles to the vertical cube diagonal. Finally, we must imagine a deformation of the rock-salt cube into a rhombohedron.[1]

A peculiar type of crystal is represented by *zinc blende* (ZnS), consisting of two interpenetrating, face-centred, cubic lattices of zinc and sulphur atoms (Fig. 50). Carborundum (SiC) and fluorspar (CaF_2) crystallize in the same way. On substituting carbon atoms for both the zinc and the sulphur atoms in zinc blende, we obtain the *diamond lattice* (Fig. 51). In this each atom represents the centre of a tetrahedron whose corners are occupied by four other carbon atoms. Silicon crystallizes like diamond.

Graphite, another modification of carbon, crystallizes

[1] $MnCO_3$, $FeCO_3$, and $NaNO_3$ crystallize like calcite.

rhombohedrally as shown in Fig. 52; and the crystals of arsenic, antimony, and bismuth are similar in structure. Numerous elements, finally, have been found to possess *hexagonal* lattices, *e.g.*, beryllium, magnesium, zinc, cadmium, titanium, zirconium, ruthenium, and osmium.[2]

§ 42. Molecular Spectra.

Although, as yet, we know nothing of the way in which the molecules of a polyatomic gas are constructed from the atoms, yet it seems at least plausible to assume that the atoms in the molecules can *oscillate* with respect to each other, and further that the molecules can execute *rotations* (apart from translatory motions). On these assumptions, the *internal energy of a molecule* would consist of three components: the internal energy of the atoms (E_{I}), the energy of the oscillatory motion (E_{II}), and the energy of the rotatory motion (E_{III}). Every *variation* in the internal energy of a molecule could therefore be represented in the form

$$(1) \qquad \Delta E = \Delta E_{I} + \Delta E_{II} + \Delta E_{II} \,.$$

If we wish to apply *Bohr's frequency condition* also to the variations of intra-molecular energy, we must assume that a variation of the energy by ΔE is connected with the emission or absorption of a light-quantum whose frequency is equal to $\Delta E/h$. From the point of view of the quantum theory, we must moreover suppose that the quantities E_{II} and E_{III} can only assume *quite definite values determined by quantum relations*, as we know already from previous considerations to be the case for the quantity E_{I}. Granted that this is so, however, the possible values of ΔE_{II} divided by h must then also represent a manifold of quite definite frequency values (ν_{II}); and, likewise, the possible values of ΔE_{III} divided by h will form another manifold of quite definite frequency values (ν_{III}). We have, in addition, the manifold of optical spectral lines (ν_{I}), due to those variations which only affect the internal atomic energy (E_{I}). Hence, on adding the frequency values of these three manifolds, we obtain the manifold of frequency values of the *molecular spectrum* in the form

$$(2) \qquad \nu = \nu_{I} + \nu_{II} + \nu_{III} \,.$$

[2] A review of crystal structures hitherto investigated, with references to the literature, is to be found in "*X-Rays and Crystal Structure*," by Sir W. H. Bragg and Prof. W. L. Bragg, and also in Ewald's book, "*Kristalle und Röntgenstrahlen*" (Berlin: Springer, 1923).

When, in particular, both E_{I} and E_{II} remain unchanged during a variation of the molecular energy, *i.e.*, when ν is equal to ν_{III}, we speak of a *pure rotation spectrum;* the question then arises as to what are the quantum relations through which the manifold of possible values of the *rotational energy* can be determined. The simplest and most obvious assumption to make is that a molecule consists of *rigidly inter-connected atoms*, so that a diatomic molecule might be compared approximately with a dumbbell.

On this assumption, the molecular *moment of inertia* (I) may be regarded as *constant*. Again, it follows from elementary formulæ of the mechanics of rigid bodies that the angular momentum (U) and the rotational energy (E_{III}) are given by

$$(3) \qquad U = w\,I\,, \quad E_{\text{III}} = \frac{1}{2} I\,w^2\,,$$

where w denotes the angular velocity. If we postulate that the *angular momentum* multiplied by $2\,\pi$ shall be an *integral multiple of the elementary quantum of action*, we obtain from eqn. (3) the relations

$$(4) \qquad w = \frac{n\,h}{2\,\pi\,I}\,,$$

and

$$(5) \qquad E_{\text{III}} = \frac{n^2\,h^2}{8\,\pi^2\,I}\,,$$

where n invariably denotes a whole number—a quantum number. Hence, whereas the energy of an electron in an atom is inversely proportional to the square of its quantum number, the rotational energy of the molecules is *directly* proportional to the square of the quantum number.

If n and s be two quantum numbers, we thus have

$$(6) \qquad \nu_{\text{III}} = \frac{h}{8\,\pi^2 I}\,(n^2 - s^2)\,.$$

Now, it may also be shown by the same methods of reasoning as led to the selection principle of Rubinowicz (*vide* § 12) that the quantum number of the rotational energy can only vary by *unity*.[1] We have therefore

$$(7) \qquad n^2 - s^2 = \pm\, 2n - 1\,.$$

[1] The numbers n and s must, of course, *not* be identified with the *principal* quantum number of earlier considerations.

Putting for brevity

(8) $$\frac{h}{8\pi^2 I} = \nu^*,$$

we thus obtain

(9) $$\nu_{III} = \nu^* (\pm 2n - 1).$$

The frequencies of the individual lines in a pure rotation spectrum must therefore be *odd integral multiples* of the fundamental frequency defined by eqn. (8). An absorption spectrum possessing this property has actually been detected in some elements *far in the infra-red*, notably by *Rubens* and *Eva von Bahr* in *water vapour*.[2]

A pure rotation spectrum can, however, be *superimposed* upon one of frequency ν_{II} arising from a variation of the oscillatory energy. In that case, provided ν_{II} be considerably greater than ν^*, there appears a so-called *rotation-*

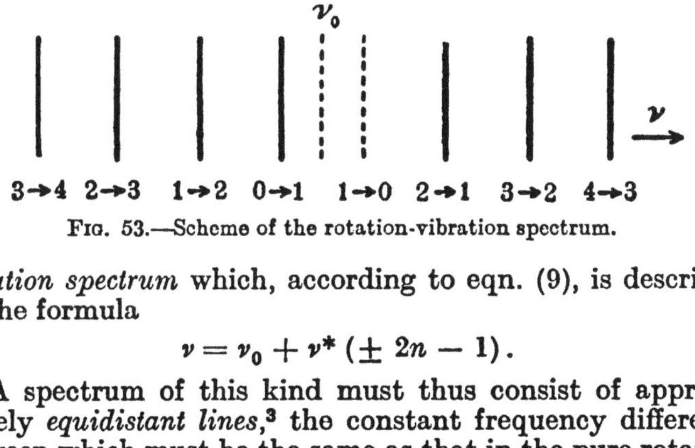

FIG. 53.—Scheme of the rotation-vibration spectrum.

vibration spectrum which, according to eqn. (9), is described by the formula

(10) $$\nu = \nu_0 + \nu^* (\pm 2n - 1).$$

A spectrum of this kind must thus consist of approximately *equidistant lines*,[3] the constant frequency difference between which must be the same as that in the pure rotation spectrum. The scheme of such a spectrum is represented in Fig. 53; underneath the individual lines are given the values of the numbers n and s corresponding to the transition $n \to s$, whereby, in accordance with the above-mentioned selection principle, s can only differ from n by unity.

Series of *absorption lines* not far in the infra-red (at wavelengths of a few μ), corresponding to this scheme, have

[2] Rubens, *Berlin. Ber.*, 1913, p. 513; and Bahr, *Verh. der Deutsch. Physikal. Gesellsch.*, **15** (1913), pp. 731 and 1150.

[3] The equidistance is only approximate, because the intervals between the lines depend not directly on the frequency differences, but on differences between angles of deflection.

actually been detected and investigated in many substances, especially in the hydrogen compounds of the halogens (HCl, HF, HBr).[4] Several series of lines occur in the various substances, which is due to the fact that ν_0 may also be a so-called harmonic. In Fig. 54 is reproduced a photograph by *Imes* [5] of a series of absorption lines, a so-called *absorption band* belonging to hydrochloric acid and lying between

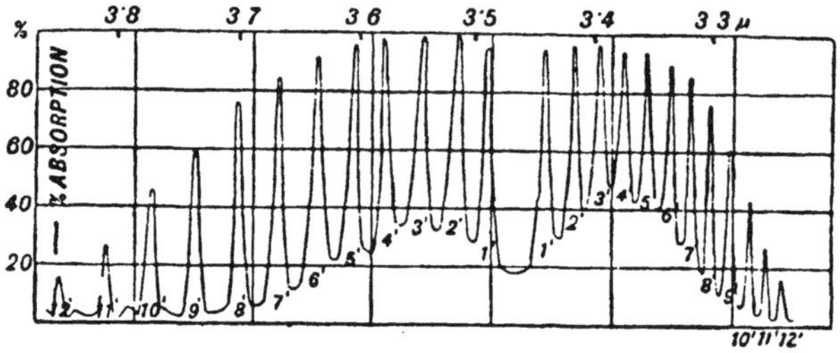

Fig. 54.—Absorption bands in HCl.

3 and 4 μ (30,000 and 40,000 Å.U.). The ordinates give the absorption expressed as a percentage, and the abscissæ angles of deflection measured in connection with the photograph, which was taken with a reflection grating. The maxima along the curve represent the absorption lines, and the corresponding wave-lengths (in μ) are given above them; below the maxima are inserted the relevant values of the

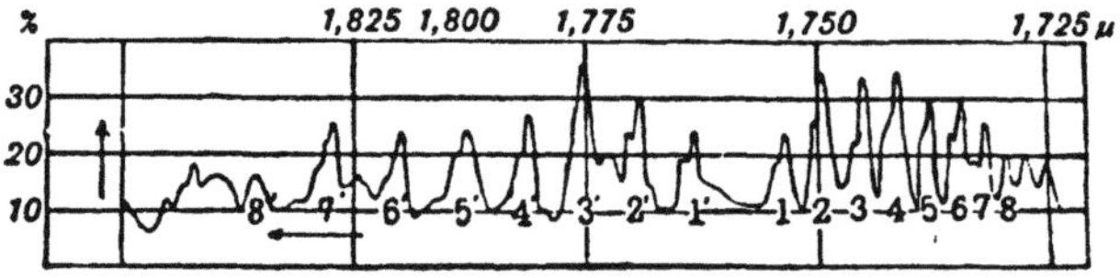

Fig. 55.—Isotope effect in the rotation-vibration spectrum of HCl.

number n following from eqn. (10). The form of the curve and a comparison with Fig. 53 (in which the transitions are given for the case of emission) show that the state for which $n = 0$ is not represented as an *initial state* in the *absorption spectrum*. We must therefore assume that a non-excited molecule *never*, or only *very rarely*, exists in a *rotationless* state.

[4] The experimental investigation of rotation-vibration spectra is principally due to Rubens, Eva von Bahr, Sleator, and Imes.

[5] *Astrophys. Journ.*, 50 (1919), p. 251.

The value of ν_0, the so-called "vibration frequency," evidently depends on the *mass* of the oscillating atom,[6] and it is thus to be expected that *isotope doublets* (or multiple lines) will occur in the rotation-vibration spectra of compounds of mixed elements.[7] Doublets of this kind, due to the isotopes of *chlorine*, are in fact clearly recognizable in Fig. 55, in which is reproduced an absorption band of HCl at 1·76 μ, according to Imes.[8] In the neighbourhood of the principal maxima due to the Cl_{35} atoms, the curve exhibits weaker subordinate maxima of smaller frequency, occasioned by the Cl_{37} atoms.[9]

Inasmuch as the values both of ν_{II} and of ν_{III} lie in the infra-red, they are relatively small as compared with optical frequencies. If, therefore, changes in the molecular energy be accompanied by a *variation in the intra-atomic energy* (E_I)

FIG. 56.—The so-called cyanogen bands in the nitrogen spectrum.

which, when divided by h, gives an *optical* frequency, a series of closely packed lines will be produced in the optical region of the spectrum, forming a so-called *band*. The positions at which the lines appear particularly closely packed are called the *heads of the bands*. The whole assembly of bands forms the so-called *band spectrum* of the molecule concerned. By way of illustration, a portion of the band spectrum of *nitrogen* (the so-called cyanogen bands) is reproduced in

[6] For ν_0 corresponds to the variation in oscillatory energy of the atoms combined in the molecule.

[7] The existence of isotopes also finds expression in a difference in the molecular moments of inertia.

[8] This band corresponds to a "harmonic," for the fundamental frequency in Fig. 55 is double that in Fig. 54.

[9] The difference in frequency between the two lines of an isotope doublet of Cl is theoretically given by

$$\frac{\delta\nu}{\nu} = \frac{1}{2}\left(\frac{1}{35} - \frac{1}{37}\right).$$

It follows that the wave-length difference of the doublet in the band reproduced in Fig. 55 is 13 Å.U.

Fig. 56; the numbers underneath the figure give the wave-lengths in Å.U.[10]

The theory of band spectra (into which we cannot enter here)[11] is very complex, since the electron transitions generating the optical frequencies (ν_1) will obviously also produce variations in the molecular moment of inertia. The most important laws of band spectra were discovered empirically by *Deslandres* in 1887.[12] An explanation of these spectra from the point of view of the quantum theory was first attempted by *Bjerrum* in 1912, or before the birth of Bohr's theory of the atom.[13] In 1916 *Schwarzschild* applied Bohr's frequency condition and the principle of quantizing the angular momentum to molecular spectra, and since then the theory has been developed principally by *Heurlinger*, *Lenz*, and *Kratzer*, although no real finality has as yet been reached.

§ 43. Dispersion.

In consequence of the revolutions of the electrons, the oscillations of the atoms, and the rotations of the molecules, numerous "*characteristic frequencies*" are to be observed in the molecules of all substances; these frequencies lie in the infra-red, in the optical region of the spectrum, and in the ultra-violet. Hence, when a molecule is penetrated by a light-wave of a frequency which does not differ greatly from a characteristic frequency, a kind of *resonance* ensues, and to this is to be ascribed the well-known phenomenon of *dispersion*, discovered by *Newton* in 1672.

It has not yet been possible to provide a satisfactory explanation of dispersion from the point of view of the

[10] The *many-line spectrum of the hydrogen molecule* is to be included among band spectra. The absence of band heads in it can be explained by the exceptional smallness of the moment of inertia of the hydrogen molecule. Helium also shows a many-line spectrum, which must be referred to a transitory formation of molecules: *cf.* the remarks in § 38 on metastable helium atoms.

[11] *Cf.* Sommerfeld's "*Atomic Structure and Spectral Lines*," Ch. vii, §§ 3–5; and A. Kratzer's "*Der heutige Stand der Theorie der Bandenspektren*," in "*Ergebnisse der exakten Naturwissenschaften*," Vol. I.

[12] The most important regularity discovered by Deslandres is to the effect that, with very fair approximation, the individual lines in a band may be represented in the form

$$\nu = \nu_1 + an^2,$$

where ν_1 is the frequency of the head of the band, a a constant characteristic of the band in question, and n successively assumes all integral values.

[13] Bjerrum, it is true, quantized the rotational energy instead of the angular momentum, and, also incorrectly, equated the frequencies of the spectral lines simply to the revolution numbers of the rotation.

quantum theory.[1] The older theory, which assumed quasi-elastic forces between the electrons in an atom, led to a dispersion formula that agrees well with experimental results. This formula [2] gives the following expression for the square of the *refractive index* :—

$$(1) \qquad n^2 = 1 + \sum \frac{N_i e^2 z_i}{\pi m_i (\nu_{0i}^2 - \nu^2)} .$$

In this equation, different kinds of electrical charges are assumed to exist and they are distinguished by the subscript i ; $z_i e$ is the quantity of electricity in such a charge, m_i its mass, ν_{0i} the frequency of its characteristic oscillation and N_i the number of these charges in a unit-volume.

Eqn. (1) shows in the first place that, with increasing values of ν, the individual summands likewise increase, both when $\nu > \nu_{0i}$ and when $\nu < \nu_{0i}$. (For, in the first case, the summand is positive and its denominator decreases, and, in the second case, the summand is negative and its denominator increases.) This is actually confirmed by experience, for, in almost all substances, the refractive index in the visible region of the spectrum *increases from the red towards the violet end.*

On the other hand, a very different behaviour must be expected when the frequency of the light differs but little from a characteristic frequency of the substance. In accordance with eqn. (1), as the increasing value of ν approaches the value ν_{0i}, the summand in question becomes larger and larger, and would finally, if eqn. (1) were actually obeyed exactly, become infinite. As ν increases beyond ν_{0i}, it follows from eqn. (1) that n^2 must first become negative, *i.e.*, the refractive index would be imaginary. It is not until ν becomes larger than ν_{0i} by a definite amount that n will again become real, and, starting from zero, again increase.

In the region of a continuous spectrum near the frequency of a characteristic oscillation, light with a frequency larger than ν_{0i} will therefore be less refracted than light with a frequency smaller than ν_{0i}. This is, of course, a contradiction to the rule, generally confirmed by experience, that the refractive index increases with the frequency. If, therefore,

[1] *Cf.* Ladenburg and Reiche, "*Absorption, Zerstreuung und Dispersion in der Bohrschen Atomtheorie,*" *Naturwissenschaften*, **11** (1923), p. 584.

[2] If, in particular, we equate to zero the frequency ν in eqn. (1), we obtain the *dielectric constant* in the narrower sense, characterizing the electro*static* field. The formula thus specialized gives a description of the *deviations from Maxwell's relation.* This relation is obeyed exactly only in such wave-length regions as contain no characteristic oscillations of the substance in question.

a characteristic oscillation of a substance falls within the *visible region* of the spectrum, the successive order of colours in the latter may differ from the usual one. An *anomalous dispersion* of this nature is actually observed in some substances, notably in *iodine vapour* and *fuchsine*. Unlike the older experimental physics, however, the electron theory does not discriminate between normal and anomalously dispersing substances. Indeed, according to the electron theory, *every* substance should exhibit the phenomena both of normal and of anomalous dispersion. If the dispersion in the visible part of the spectrum be "normal" for most substances, the electron theory only draws the conclusion that the characteristic oscillations responsible for anomalous colour dispersion lie *outside* the visible region of the spectrum in the majority of substances.

Waves whose frequencies differ very little from a characteristic frequency are very strongly reflected and absorbed by the substance in question. This phenomenon, which has been observed in many substances, is known as *selective reflection* and *absorption*. Observations on the phenomenon render possible a determination of the values of the characteristic frequencies, and, with the help of the dispersion formula, they also give clues as to the nature and the number of the charges responsible for the dispersion. Indeed, as might be expected, observations have shown that characteristic frequencies in the ultra-violet arise from valency electrons, whereas those in the infra-red, which in many substances extend into the region of short electric discharge waves, are to be referred to atomic oscillations.

§ 44. The Radiation Constants.

Before the birth of the quantum theory, researches on *thermal radiation* had led to the discovery of *two universal constants*, the atomistic signification of which was rendered manifest by the quantum theory; these two constants are known as *Stefan's* and *Wien's* constant respectively.

In 1879, it was found by Stefan that the amount of energy radiated from unit-area of a black body[1] in unit-time, *i.e.*, the *emissive power* of the black body, is *proportional to the fourth power of the absolute temperature*.[2] The emissive

[1] A body is said to be perfectly black when it absorbs completely an incident electromagnetic wave radiation.

[2] This law is frequently called the Stefan-Boltzmann law, for, whereas Stefan's formulation of it was only empirical, Boltzman was the first to derive it theoretically, on the basis of thermodynamical considerations.

power may be equated to σT^4, where the value of σ (Stefan's constant), as found by experiment, is given by

(1) $$\sigma = 5{\cdot}76 \times 10^{-5};$$

the radiated energy is measured in ergs, and the temperature in degrees Centigrade.

Again, it was discovered by Wien in 1893 that, in the continuous spectrum of a radiating body, the *wave-length* corresponding to the *maximum specific emissive power* is *inversely proportional to the absolute temperature.*[3] The product of this wave-length and the temperature represents Wien's constant, for which measurements have given a value of

(2) $$b = 0{\cdot}288 \text{ cm. deg.}$$

Now, in 1900, *Planck* showed that both constants are directly connected with the elementary quantum of action and the Loschmidt number (we cannot, however, discuss the proof here), and he was thus able to calculate both these quantities from Stefan's and Wien's constants. In this way he obtained the first exact value for the Loschmidt number of which only the order of magnitude had previously been known, and, at the same time, he derived also an exact value for the elementary quantum of action, which he was the first to introduce into physics.[4] For, as Planck found,

(3) $$\sigma = \frac{2\pi^5}{15\, h^3 c^2}\left(\frac{R}{L}\right)^4,$$

while

(4) $$b = \frac{chL}{\beta R}$$

where $R(= 8{\cdot}313 \times 10^7$ erg./deg.) denotes the absolute *gas constant* [5] and β is the pure number 4·9651;[6] L, h, and c

[3] This law is generally termed Wien's displacement law, because, on a change in temperature, the position of maximum emissive power in the spectrum is displaced.

[4] Planck obtained for L a value of $6{\cdot}17 \times 10^{23}$, and for h a value of $6{\cdot}548 \times 10^{-27}$ erg. sec. These values agree extraordinarily well with those resulting from later and much more accurate investigations of a different kind. From the Loschmidt number and the known value of Faraday's electrochemical constant, Planck derived, in 1900, a value of $4{\cdot}69 \times 10^{-10}$ E.S.U. for the elementary quantum of electricity.

[5] The gas constant, multiplied by the absolute temperature, represents the product of the volume and the pressure of a gram-molecule of a perfect gas.

[6] β is the root of the transcendental equation

$$\frac{\beta}{5} + e^{-\beta} = 1.$$

denote, as usual, the Loschmidt number, the elementary quantum of action, and the velocity of light. The quantities L and h may be regarded as unknowns in the two eqns. (3) and (4), and the equations may thus be solved with respect to them.

We cannot enter into details here, but a direct and exact determination of the Loschmidt number can also be effected by means of observations on the so-called *phenomena of molecular fluctuations*.[7] Thus, exact measurements of the *Brownian movement*[8] have given a value of $5{\cdot}94 \times 10^{23}$ for the Loschmidt number,[9] and the value found from determinations of the concentrations of *emulsions*[10] at different heights[11] is $6{\cdot}04 \times 10^{23}$.

Modern physics thus affords at least seventeen different and independent equations for the four fundamental constants of atomistics, *viz.*, h, e, L, and m (mass of an electron). These equations contain only these constants as unknowns in various combinations, as may be seen from the summary in Table XVII.

In whatever way the four unknowns are derived from these seventeen equations, the same values within the limits of error are always obtained. If the fact that the sensations

[7] The phenomena of molecular fluctuations originate in the general fact that an average value, which is obtained from only a small number of individual values, may experience relatively considerable variations, not through any external cause, but simply because of individual irregularities. Spontaneous fluctuations must therefore be perceptible in all physical quantities, the value of which depends on only a small number of individual values. Conversely, the observation of such fluctuational phenomena may be regarded as a certain proof of the individual composition of matter and physical processes.

[8] The Brownian movement, which was discovered in 1827, originates in the irregular molecular impacts experienced by a small body suspended in a gas or a liquid. Since even a small body experiences millions of impacts in a minute fraction of a second, the effects of these will cancel out in consequence of the statistical regularity due to their large number, inasmuch as they take place in all directions. The smaller the suspended body, however, the greater is the part played by individual accidental irregularities. The mutual compensation of the impacts occurring in all directions becomes no longer perfect, and, if the body be sufficiently small, it will be forced into a zig-zag motion which will be the more lively the smaller the body.

[9] Erich Schmid, *Sitz. Ber. der Wien. Akad.*, 1921.

[10] Emulsions are distinguished from solutions by the fact that, in them, matter is not resolved into molecules but is suspended as particles in the liquid. These particles are composed of many molecules and can be observed in a microscope, and their size can also be measured. Such emulsions are also known as colloidal solutions. Under the influence of gravity, a definite distribution is formed in emulsions, and to this is due the variation of the concentration with the height.

[11] Westgren, *Zeitschr. f. anorg. Chem.*, **95** (1915), p. 231.

of sight, hearing, and taste all indicate the same object is to be taken as an argument for existence of the world around us, then theoretical physics may regard as a *certain proof of the actual existence of electrons, atoms, and the elementary quantum of action* the fact that the same values are always obtained for the characteristic constants of atomistics, even by fundamentally different methods.

TABLE XVII

Methods of Derivation of the Fundamental Constants of Atomistics

Method.	Reference.	Relevant combination between e, h, m, and L,
1. Individual observation of small electric charges.	§ 1 eqn. (7)	e
2. Electrolysis . . .	§ 2 eqn. (5)	$L\,e$
3. Deflection of cathode rays .	§ 3 eqn. (4)	e/m
4. Deflection of α-rays .	§ 4 eqn. (1)	$L\,e$
5. Scintillation of α-particles .	§ 4	e
6. Photo-electric effect .	§ 5 eqn. (2)	e/h
7. Rydberg's constant . .	§ 9 eqn. (8)	$e^4 m/h^3$
8. Displacement towards the violet of helium lines	§ 11 eqn. (25)	Lm
9. Fine structure of spectral lines.	§ 13 eqn. (3)	e^2/h
10. Stark effect . . .	§ 14 eqn. (2)	h/em
11. Zeeman effect . . .	§ 15 eqn. (3)	e/m
12. Excitation of Röntgen rays	§ 20 note (4)	$e^{4/3}/h$
13. Formation of helium from radium.	§ 24	L
14. Stefan's Law . . .	§ 44 eqn. (3)	$h^3 L^4$
15. Wien's Law . . .	§ 44 eqn. (4)	$L\,h$
16. Brownian Movement .	§ 44	L
17. Emulsions . . .	§ 44	L

APPENDIX

SUMMARY OF THE CONTENTS

Chapter I

The Elementary Quanta.

§ 1. It has been found from the movements executed by small particles, with weak electrical charges, under the dual influence of their own weight and an electric field directed vertically upwards that the charges on such particles are exact integral multiples of an elementary quantum of electricity (e). The value of this quantum is found by observation to be $4{\cdot}77 \times 10^{-10}$ E.S.U.

§ 2. The number by which the atomic weight numbers must be divided in order to obtain the absolute masses of the atoms is termed the Loschmidt number (L). The product Le is known from electrolytic measurements, on the basis of Faraday's fundamental laws, and the value of L thus follows as $6{\cdot}06 \times 10^{23}$. Hence the mass of an atom of hydrogen is $1{\cdot}66 \times 10^{-24}$ gm.

§ 3. The specific charge of cathode ray particles is found from their electric and magnetic deflections to be about 1,800 times that of an ionized atom of hydrogen. On the assumption that the charge on each particle is one elementary quantum, the cathode rays thus consisting of electrons, the mass (m) of an electron is found to be $9{\cdot}0 \times 10^{-28}$ gm. The velocity of cathode rays is proportional to the square root of the tube potential. The negative corpuscular rays emitted by radioactive substances, or the so-called β-rays, likewise consist of electrons, and their velocities amount to from 30 to 99·8% of the velocity of light. It has been possible to demonstrate experimentally with these rays the increase in mass with the velocity demanded by the theory of relativity.

§ 4. Canal rays and anode rays consist of positive ions, whose masses are found by positive ray analysis to be always approximately integral multiples of the mass of a hydrogen atom. The positive α-rays emitted by radioactive substances have initial velocities of 5—7% of the velocity of light. The scintillation method allows of a count of the α-particles expelled by a preparation in a given time. In this way the charge on a single α-particle can be derived, and hence also its mass, since the specific charge is known from deflection experiments. It has been found that the mass of an α-particle is four times as large as that of a hydrogen atom, whilst its positive charge amounts to two elementary quanta. In hydrogen, α-particles produce long-range H-rays.

§ 5. Careful investigations on the photo-electric effect show that the equivalent potential of the liberated electrons is a linear function of the frequency of the radiation producing them. This leads to the existence of a universal constant with the dimensions of action, or of an elementary quantum of action (h); its value is found from measurements to be $6{\cdot}55 \times 10^{-27}$ erg. sec. According to a law established by Einstein, every mutual transformation between corpuscular radiation and electromagnetic wave radiation takes place in such a way that, either a single light-quantum ($h\nu$) is used to liberate a single electron, or, in the reverse case, a single light-quantum is produced by the energy of a single electron.

Chapter II

Theory of the Hydrogen Atom.

§ 6. Observations on the passage of α-rays through matter show that an α-particle can penetrate through several thousands of atoms without experiencing any appreciable change in direction, although sometimes a single atom produces a very large deflection of the α-particle. The deflection must therefore be due to an atomic nucleus which occupies only a very small part of the total volume, and in which almost the whole mass of the atom is concentrated. According to Rutherford's fundamental conception, the nucleus must be regarded as positive and as surrounded by revolving electrons, the electrical force of attraction exerted on these being balanced by the centrifugal force.

§ 7. Bohr's fundamental hypothesis is to the effect that the angular momentum of the normal hydrogen atom multiplied by 2π shall be equal to the elementary quantum of action, but, besides the normal state, other higher quantum states shall also be possible, in which the angular momentum is an integral multiple of the normal. In any arbitrary state, the energy, which is always negative, is equal to the normal energy divided by the square of the quantum number.

§ 8. On the assumption that every transition between two different states compatible with the quantum theory is accompanied by the emission or the absorption of a light-quantum (according as the quantum number is lowered or raised), Bohr obtained for the hydrogen spectrum the formula

$$\nu = R\left(\frac{1}{s^2} - \frac{1}{n^2}\right), \qquad R = \frac{2\pi^2 e^4 m}{h^3}.$$

The frequencies of the lines thus appear representable as differences between terms.

§ 9. The spectral formula theoretically derived by Bohr agrees with that experimentally obtained by Balmer. According as the quantum number s, which corresponds to the final state for emission, is given the value 1, 2, 3, or 4 in Bohr's formula, an expression is obtained for the ultra-violet Lyman series, the optical Balmer series, the infra-red Paschen series, or the Brackett series of long wave-length.

§ 10. If the frequency in Bohr's formula for the hydrogen spectrum be multiplied by 2^2 the expression thus obtained must theoretically represent the spectrum of ionized helium. In point of fact these series, which were originally erroneously ascribed to hydrogen, have been shown by later experiments to be due to helium.

§ 11. The displacement towards the violet of the helium lines relatively to the hydrogen lines is explained, as Bohr recognized, by the simultaneous motion of the atomic nucleus. Measurements on the displacement towards the violet yield an exact determination of the specific charge of an electron.

§ 12. In Sommerfeld's development of Bohr's theory, each of the electronic orbits regarded as ellipses requires two quantum numbers for its definition. The principal quantum number determines the major axis of the ellipse, and the subordinate number its eccentricity. The selection principle permits only of those transitions between orbits in which the subordinate number changes by unity.

§ 13. Although, as a first approximation, the energy of the electron in the hydrogen atom is independent of the subordinate number of its orbit, a dependency is found to exist within narrow limits when, in accordance with the theory of relativity, the change of mass with velocity is taken into account. In this way Sommerfeld succeeded in explaining the fine structure of the hydrogen lines; his theory is completely confirmed by experiment.

§ 14. The resolution of spectral lines in an electric field, or the so-called Stark effect, may be regarded as a brilliant experimental confirmation of the quantum theory.

§ 15. Just as in an electric field, so also in a magnetic field there occurs a resolution of the spectral lines proportional to the field-strength: this is known as the Zeeman effect.

CHAPTER III

Röntgen Rays.

§ 16. Röntgen rays are generated by the arrest of electrons moving with high velocities.

§ 17. A space-lattice separates by diffraction rays of quite definite wave-lengths from out of an incident wave radiation. These wave-lengths depend, for given ordinal numbers, on the lattice-constant and the direction of incidence.

§ 18. The diffraction of rays by a space-lattice may also be regarded as a reflection at the lattice-planes. A set of lattice-planes singles out from an incident radiation those rays for which an integral multiple of the wave-length is equal to the difference in path between two rays which are reflected from two successive lattice-planes. On rotation of a space-lattice, all rays of the same wave-lengths are focussed at definite points lying on the circumference of a circle.

§ 19. The diffraction of Röntgen rays by crystals was demonstrated by Laue. The investigation of the structure of a crystal on the basis of the Laue photographs rendered possible an exact measurement of the wave-lengths of Röntgen rays. By means of crystals, wave-lengths of from about 100 to 17,000 X.U. can be measured.

§ 20. Duane and Hunt's equation, which follows from Einstein's law of light-quanta, states that

$$\lambda V = 1{\cdot}234 \times 10^7 \qquad (\lambda \text{ in X.U.}, V \text{ in volts.}).$$

In complete agreement with experimental results, this equation determines, on the one hand, the excitation potential for a given wave-length, and on the other hand, the smallest possible wave-length for a given tube potential, in each case in the continuous Röntgen spectrum.

CHAPTER IV

Theory of the Elements.

§ 21. The foundation of Röntgen spectroscopy by Moseley is based on Barkla's discovery of characteristic radiations. As Moseley found, the elements can be arranged in a consecutively numbered series, in such a way that the square root of the frequency of a definite Röntgen line increases linearly with the number (the atomic number) of the element. The " natural sequence " of the elements thus obtained, which begins with hydrogen and ends with uranium, comprises ninety-two places. The atomic number proves to be identical with the nuclear charge number.

§ 22. The natural sequence of the elements is divided by the chemically inert rare gases into seven periods. The first period contains 2 elements, the second and third periods contain 8 elements each, the fourth and fifth 18 each, the sixth contains 32 elements, and the seventh period breaks off at the sixth place with uranium. Homology among the elements allows of a distinction between eight groups, each with two sub-groups, in the periodic system.

§ 23. The phenomenon of isotopy, *viz.*, the chemical identity of substances with different radioactive properties or atomic weights, is explained by the fact that two atoms with the same nuclear charge may yet possess a different nuclear structure, and thus a different number of protons. The analysis of mixed elements was rendered practicable by positive ray analysis, which has been developed by Aston into mass-spectroscopy.

§ 24. The phenomena of radioactivity are due to a disintegration of the atomic nuclei, which, owing to the alteration in the nuclear charge number, at the same time effects a chemical transformation. An α-transformation reduces the atomic number by two, and the atomic weight by four, whereas a β-transformation raises the atomic number by one, without noticeably altering the atomic weight. The observation of the rate of formation of helium from radium, combined with the scintillation method, permits of a direct determination of the Loschmidt number.

§ 25. In radioactive equilibrium, the numbers of atoms present vary inversely as the disintegration constants. Equal numbers of atoms of all substances which are in equilibrium with each other disintegrate in the same time.

§ 26. At the present time three transformation series are known, commencing with uranium, actinium, and thorium respectively, and all ending with stable species of lead. Besides these series, potassium and rubidium are also known to be radio-elements. The age of uranium minerals can be calculated from their content of uranium lead.

§ 27. The range of α-rays is proportional to the cube of their initial velocity. The logarithms of the disintegration constants are linear functions of the logarithms of the ranges. β ray spectra are produced by the deflection of β-rays in magnetic fields. For γ-rays it has been possible to measure wave-lengths as small as 20 X.U., and to estimate wave-lengths as low as 5 X.U.

§ 28. Using α-rays, Rutherford succeeded in disrupting nitrogen nuclei, and subsequently he and other investigators have been similarly successful with 20 other elements, hydrogen nuclei being expelled from the artificially disrupted nuclei.

§ 29. In elements with an atomic weight divisible by four, the nuclei are very possibly built up solely of positive and neutralized α-particles. Elements of even atomic number are of more frequent occurrence than those of odd atomic number, and are generally of even atomic weight. Elements of odd atomic number are either pure elements, or possess two isotopes, the atomic weights of which are generally odd and differ by two. The linear dimensions of the nuclei prove to be of the order of magnitude of 10^{-12} cm. The comparatively enormous internal energy of the atomic nuclei is manifested, in accordance with the Law of the Inertia of Energy, in the deviation from whole numbers of the atomic weights referred to hydrogen.

CHAPTER V

General Theory of Spectra and Atomic Structure.

§ 30. The K-, L-, M-, and N-series of Röntgen spectra arise from transitions to a one-, two-, three-, and four-quantum final state. Since the excitation potential necessary for the generation of a definite Röntgen line is always found to be equal to the characteristic potential of the final level, a Röntgen line can obviously be produced only when an electron in the emitting atom has been previously removed from its normal orbit to beyond the limits of the atom. This hypothesis is afforded a direct experimental confirmation by the absorption edges in the continuous Röntgen spectrum.

§ 31. An explanation of the multiplicity of observed Röntgen lines demands an assumption of several modifications in the various quantum levels. According to Bohr, each level can be represented by three quantum numbers in the symbolical form S_{kj}. S represents the principal quantum number; the subordinate quantum number k cannot be greater than the principal quantum number, and the inner quantum number j is either equal to k or less by one, a distinction being thereby made between normal and anomalous levels. Only those transitions between levels are apparently possible in which k changes by unity, and j either remains unaltered or also changes by unity. The wave-lengths of the γ-rays of a radio-element can be calculated from its magnetic β-ray spectrum, provided that the energy-levels of an isotopic element are known from Röntgen-spectroscopic measurements.

§ 32. In an excited state of an atom, an electron can describe a path of higher quantum number than the normal one. Excitation may be due to the absorption of light-quanta, or, in the case of thermal luminescence, to the transformation of translational energy. A sufficient introduction of energy may lead to the ionization of atoms, to which, *inter alia*, the phenomenon of the conductivity of flames may be referred. Besides the arc spectrum as the spectrum of the neutral atom, elements also exhibit spark spectra as the spectra of their ions. According to the spectroscopic displacement law, the spark spectra of the elements in a group of the periodic system agree in character with the arc spectra of the preceding group.

§ 33. Optical absorption spectra arise on the transition of an electron to a higher quantum orbit, either from the fundamental orbit or from a previously excited orbit. Optical emission spectra arise on the spontaneous return, in either one or more stages, from unstable states to the fundamental state. The lines in optical series converge with decreasing wave-length, decreasing intensity, and decreasing sharpness towards a limit which, in the case of emission, corresponds to the final level of the transition. The most important lines can be referred to four series of terms, which are called the *s*-, *p*-, *d*-, and *f*-terms, and characterized by the subordinate quantum numbers 1, 2, 3, and 4. The principal series, the diffuse subordinate series, the sharp subordinate series, and the Bergmann series correspond, in the case of emission, to the transitions $p \rightarrow s$, $d \rightarrow p$, $s \rightarrow p$, and $f \rightarrow d$. In the arc spectra of the alkali metals, all terms are double, with the exception of the *s*-terms, which are always simple. In the arc spectra of the elements of the second group, there is a system of simple terms as well as one of triple terms, and spectroscopic combinations may occur between both systems.

§ 34. Corresponding to the scheme of Röntgen levels, the electrons revolving about an atomic nucleus may be divided up into groups and sub-groups. The number of electrons in a fully completed sub-group is equal to twice the inner quantum number of the Röntgen level corresponding to the sub-group. The electron grouping in the atoms of the rare gases may thus be derived from the Röntgen spectra. In the emanation atom, the one-quantum group contains 2, the two-quantum group 8, the three-quantum group 18, the four-quantum group 32, the five-quantum group 18, and the six-quantum group 8 electrons.

§ 35. According to Bohr's theory, the development in the periodic system of the outermost group of electrons at any time is interrupted by the completion of inner electronic groups. The completion from 8 to 18 electrons takes place in the three-quantum group between Sc and Cu, in the four-quantum group between Y and Ag, and in the five-quantum group between La and Au. Further within the atom, the four-quantum group is completed from 18 to 32 electrons throughout the rare earths between Ce and Lu, so that the 72nd element, which succeeds Lu, cannot be a rare earth, but must be a homologue of zirconium, as has actually been confirmed by the discovery of hafnium. Elements belonging to series within which an inner group of electrons is being completed frequently exhibit in consequence of the asymmetry of the inner atomic structure, the phenomena of paramagnetism, multiple valency, and the formation of coloured salts. A direct experimental confirmation of

Bohr's theory of the periodic system is afforded by the course of the curves representing the relation between the square roots of the Röntgen-spectroscopic terms and the atomic numbers.

§ 36. The radiation emitted by a body whose atoms have been excited by primary radiation is known as fluorescent radiation; it is termed resonant radiation when the secondary and primary wave-lengths are identical. Such is the case with the alkali metals when the primary wave-length is one of the doublet with the greatest wave-length in the principal series. Fluorescence in gases can be extinguished by increasing the pressure; observations made in this connection have shown that the value of the mean duration of an excited atomic state is about 10^{-8} sec. The phenomena observed of sensitized fluorescence can be referred to a transference of excitational energy by molecular collisions.

§ 37. The method of electronic collisions enables the excitation and the ionization potential to be accurately determined for many elements. The values thus experimentally obtained agree perfectly with those calculated from optical spectra on the basis of Bohr's theory.

§ 38. An excited atomic state, from which a spontaneous return to the normal state appears impossible in view of the selection principle, is termed metastable. It has been possible to produce a metastable state of helium atoms, with the help of electronic collisions.

§ 39. Magnetism in bodies is to be referred to the magnetic moments associated with the individual electronic orbits; these moments may, however, be subject to a complete or partial mutual compensation within the atoms or molecules. The reality of molecular currents has been demonstrated by Einstein and de Haas's experiment.

§ 40. Whereas the formation of positive ions is presumably due to a separation of loosely bound electrons, the formation of negative ions may be ascribed to an affinity for electrons peculiar to some elements. The combination of oppositely charged ions gives rise to heteropolar molecules, but homopolar combinations between neutral atoms and between whole molecules are evidently also possible. Through the absorption of light-quanta, molecules can be brought into a state in which, because of their increased energy, they are enabled to partake in chemical reactions not possible for them in their normal state ; photochemical phenomena are based on this absorption. According to Einstein's law of photochemical equivalence, the number of chemically reacting molecules must be equal to the number of light-quanta absorbed. The converse of photochemical phenomena is represented by chemico-luminescence.

§ 41. The formation of crystals is to be regarded as a special case of the formation of molecules. The most important types of crystal lattices are represented by rock-salt, calcite, zinc blende, diamond, and graphite.

§ 42. Molecular spectra may be explained on the assumption that the internal energy of molecules consists not only of the internal energy of the combined atoms, but also of an oscillational energy of the atoms and of a rotational energy of the molecules, and that all three kinds of energy can assume only quite definite values determined by quantum relations. Changes in the rotational energy alone give rise to rotation spectra lying in the far infra-red. Variations of both the rotational and the oscillational energy produce the oscillatory rotation spectra observed in the near infra-red. Finally, simultaneous alterations of all three kinds of energy lead to the formation of optical band spectra.

§ 43. When a molecule is penetrated by a light-wave of a frequency which does not differ greatly from a characteristic frequency of the molecule, a kind of resonance ensues, and this is manifested in the phenomena of dispersion (both normal and so-called anomalous).

§ 44. The two universal radiation constants, Stefan's and Wien's constants, are directly related to the elementary quantum of action and the Loschmidt number. It is thus possible to evaluate both these quantities from the radiation constants.

SYNOPSIS OF THE SYMBOLS MOST FREQUENTLY USED

A Atomic weight.
E Energy.
E Electric strength of field.
H Magnetic strength of field.
I Moment of inertia.
L Loschmidt number.
M Mass of a proton.
R Rydberg's constant.
T Term.
T Half value period.
V Potential.

---

a Radius.
a Lattice constant.
c Velocity of light.
d Lattice plane distance.
e Elementary quantum of electricity.
h Elementary quantum of action.
j Inner quantum number.
k Subordinate quantum number.
m Mass of an electron.
m_H Mass of the hydrogen atom.
n, s Quantum numbers.
v Velocity.
z Nuclear charge number.

α, β, γ Directional cosines.
α Ratio of velocity in one-quantum orbit in hydrogen atom to c.
γ Specific charge.
θ Glancing angle.
λ Wave-length.
λ Disintegration constant.
ν Frequency.

THE UNIVERSAL CONSTANTS OF ATOMIC PHYSICS.

e $4{\cdot}774 \times 10^{-10}$ E.S.U.
m $9{\cdot}00 \times 10^{-28}$ gm.
m_H $1{\cdot}662 \times 10^{-24}$ gm.
h $6{\cdot}545 \times 10^{-27}$ erg./sec.
c $2{\cdot}9990 \times 10^{10}$ cm./sec.
L $6{\cdot}062 \times 10^{23}$.
M/m 1846
R_∞/c $109{,}737{\cdot}1$ cm.$^{-1}$
e/m $5{\cdot}304 \times 10^{17}$ E.S.U /gm.
e/m_H $2{\cdot}871 \times 10^{14}$ E.S.U /gm.

HISTORICAL SUMMARY

c. B.C. 400. Democritus founded atomistics.
1672. Newton discovered the dispersion of light.
1805. Dalton's Law of Multiple Proportions.
1812. Berzelius referred chemical affinity to electrical forces.
1814. Discovery of the Fraunhofer lines.
1822. Ampère's theory of molecular currents.
1833. Faraday established the fundamental laws of electrolysis.
1845. Faraday's discovery that magnetism is a universal property of all substances.
1858. Plücker discovered cathode rays.
1860. Kirchhoff and Bunsen founded spectroscopy.
1865. Loschmidt first estimated the true size of molecules.
1869. Hittorf's discovery of the deflection of cathode rays in a magnetic field.
1869. Lothar Meyer and Mendeléeff discovered chemical periodicity.
1885. Balmer's spectroscopic formula.
1886. Goldstein's discovery of canal rays.
1888. Hallwachs discovered the photo-electric effect.
1895. Lorentz founded the electron theory.
1895. Discovery of Röntgen rays.
1896. Becquerel discovered radioactivity.
1896. Discovery of the Zeeman effect.
1897. J. J. Thomson and Kaufmann determined the velocity and specific charge of cathode rays.
1898. Discovery of radium by M. and Mme. Curie.
1900. Planck founded the quantum theory.
1900. Rutherford and Dorn discovered the emanations.
1902. Rutherford and Soddy founded the theory of atomic disintegration.
1903. Ramsay and Soddy detected the formation of helium from emanation.
1905. Einstein introduced the conception of light-quanta.
1905. Discovery of characteristic radiation by Barkla.

1910. Exact determination of the elementary quantum of electricity by Millikan.
1910. Soddy introduced the conception of isotopes.
1911. Rutherford introduced the conception of the nuclear atom.
1912. Laue's discovery of interference of Röntgen rays in crystals.
1913. Bohr founded the quantum theory of spectra and of atomic structure.
1913. Bohr's explanation of the displacement towards the violet of the helium lines.
1913. Moseley founded Röntgen-spectroscopy.
1913. Van den Broek recognized the identity of nuclear charge number and atomic number.
1913. W. H. and W. L. Bragg established the structure of crystals.
1913. Discovery of the Stark effect.
1913. Soddy and Fajans formulated the displacement laws governing the transformation of elements.
1913. J. J. Thomson discovered isotopy in neon.
1913. Franck and Hertz introduced the method of electron collisions.
1915. Sommerfeld introduced subordinate quantum numbers into the quantum theory.
1915. Sommerfeld's explanation of the fine structure of the hydrogen spectrum.
1915. Einstein and de Haas effected an experimental proof of the existence of molecular currents.
1916. Schwarzschild and Epstein developed the quantum theory of the Stark effect.
1916. Sommerfeld established the law of regular Röntgen doublets.
1918. Rubinowicz formulated his selection principle.
1919. Aston founded mass-spectroscopy.
1919. Rutherford effected the artificial disruption of the nuclei of elements.
1920. G. Hertz discovered the law of irregular Röntgen doublets.
1921. Bohr deduced the grouping of the electrons in the rare gases.
1922. Bohr's theory of the periodic system.
1924. Discovery of the Compton effect.

NAME INDEX

(*The numbers refer to pages.*)

SUBJECT INDEX

(The numbers refer to pages.)

PRINTED IN GREAT BRITAIN BY THE WHITEFRIARS PRESS, LTD., LONDON AND TONBRIDGE.

Bei Fragen zur Produktsicherheit wenden Sie sich bitte an:
If you have any questions regarding product safety,
please contact:

Walter de Gruyter GmbH
Genthiner Straße 13
10785 Berlin
productsafety@degruyterbrill.com